Communications in Law Enforcement

Third Edition

Communications in Law Enforcement

Third Edition

Silvana Turpin
Sault College of Applied Arts and Technology

PEARSON
Prentice
Hall

Toronto

Library and Archives Canada Cataloguing in Publication

Turpin, Silvana, 1964–
    Communications in law enforcement / Silvana Turpin. — 3rd ed.

Includes index.
ISBN 0-13-196920-X

    1. Communication in police administration. 2. Police reports.
3. Police—Records and correspondence. I. Title.

HV7936.C79T87 2006   808'.066363   C2005-900308-1

ISBN 0-13-196920-X

Vice President, Editorial Director: Michael J. Young
Acquisitions Editor: Ky Pruesse
Executive Marketing Manager: Judith Allen
Developmental Editor: Patti Altridge
Production Editor: Kevin Leung
Copy Editor: Joe Zingrone
Proofreader: Kathleen E. Richards
Production Coordinator: Janis Raisen
Manufacturing Coordinator: Susan Johnson
Composition: Christine Velakis
Art Director: Mary Opper
Cover Design: Michelle Bellemare
Cover Images: Photodisk, Digital Vision, Comstock, Firstlight

15 16 17 CP 15 14 13

Printed and bound in Canada.

Contents

Chapter 3

Letter Writing   39

Chapter 4

Notebooks   79

Chapter 5

Police Report Writing   91

Chapter 6

Public Speaking   129

Preface

Excellent written and oral communications skills are essential in the law enforcement profession. Not a shift will go by without an officer having to write a report, defend a charge in court, or talk to people to gather facts regarding an incident.

Communications in Law Enforcement has been designed to give you, the student, the knowledge and practice needed to develop the written and oral communication skills necessary for a successful career in policing. Its straightforward, step-by-step approach will aid you as you learn the theory and apply the skills discussed throughout the book. This third edition has been designed in a workbook format to give you the opportunity to work through a selection of activities.

Each chapter begins with a list of learning outcomes, then moves on to discuss the theory specific to the section in question. Numerous practice exercises are provided so that you can make use of the theory discussed. End-of-chapter activities and discussion questions allow you to review and hone your skills.

The book discusses the general concepts of the communication process, the barriers that interfere with communication, and ways to overcome these barriers. It also covers memo-, email-, and letter-writing skills that police officers use, perhaps not on a daily basis, but frequently. The third edition covers email usage more in-depth, and gives you the chance to improve on your skills and learn about email etiquette.

This edition also offers an increased focus on police report writing. Report templates used by many police services in Ontario are provided. This will allow you to see and work with the actual reports that you may encounter once you are hired by a police service as a constable. Along with the third edition of the book, you also have access to these report templates online, thus allowing you to download and submit the finished reports to your instructor. Many scenarios are provided to give you practice in interviewing, tips on taking notes in your notebook, and help with writing police narratives. Copies of the three main police reports — the general occurrence report, supplemental report, and arrest report — are provided at the end of Chapter 5, as well as six other sample police templates provided in the Appendix, to give you plenty of practise using the forms. For additional forms and to download these reports, please see the updated text enrichment site at www.pearsoned.ca/text/turpin.

Also provided in the report-writing chapter are scrambled police narratives. Working with these scenarios will help you prepare for writing-competency tests you may encounter when applying for policing positions.

You will receive guidance on how to become a concise oral communicator and afacilitate effective workshops. These are skills you will use throughout your policing career.

The book closes with a chapter on sentence skills. Any piece of writing you create should be free of grammatical errors. It is important that you review this chapter carefully and apply the rules you learn to your own writing.

Overall, this book is designed to teach you the communications skills you will need once you enter a career in law enforcement. I hope you enjoy working through it and find it helpful.

INSTRUCTOR SUPPLEMENTS

Instructors will be able to make use of the online Instructor's Manual featuring classroom quizzes and PowerPoint Presentations for *Communications in Law Enforcement*, Third Edition.

STUDENT SUPPLEMENT

A **new** Text Enrichment Site (TES) is available for students at www.pearsoned.ca/ text/turpin. This site provides students with the answers to the exercises in Chapter 8 of the book. The TES also gives students access to numerous police report templates, including the general occurrence report, supplemental report, and arrest report that are shown in Chapter 5 and the six others that are in the Appendix.

Acknowledgments

I would like to thank Ray Pritchard, Dianna McAleer, Michael Hart, Andrew M. Stracuzzi, and Halynka Honczarenko for taking the time to review the textbook and for giving suggestions. Their comments and insights are very much appreciated.

Very special thanks also go to Deputy Chief Alan Williams of the North Bay Police Service and Deputy Chief Bob Kates of the Sault Ste. Marie Police Service for their help in providing templates of the police reports. These reports, located in the text and on the TES, are used by 41 police services in Ontario. Please see the website for a complete listing of the OPTIC services that use these reports.

—Silvana Turpin

Chapter 1

The Importance of Communication in Policing

LEARNING OUTCOMES

Upon completion of this chapter, you should be able to

1. define the communication process
2. define the barriers to the communication process
3. recognize ways to overcome barriers to communication
4. recognize the importance of communication skills in policing
5. recognize the various people police officers communicate with routinely
6. define the various ways a police officer communicates routinely

THE COMMUNICATION PROCESS

Every communication process involves at least two people. The process begins with wanting to relay a message. The person sending the message is referred to as the **sender**. The person on the other end who receives the message is referred to as the **receiver**. Before the sender actually sends the message, she needs to put the message into words. The sender then sends the words through various channels. The channels used may include face-to-face contact or interaction, telephone, mail, email, faxes, memos, letters, and reports. The receiver receives the words and translates them into ideas. In most situations, the positions of sender and receiver alternate. After the receiver receives the message from the sender, he or she will want to give some feedback. The sender will then become the receiver, and the receiver will become the sender. Positions are exchanged.

BARRIERS TO COMMUNICATION

During the communication process, various barriers may disrupt the process. These barriers can lead to miscommunication because they block the process of communication and prevent the message from accurately getting to the receiver. Barriers can be classified into five major areas. These include

1. Differences in Perception

People have different beliefs formed by culture, environment, and experience. We all perceive things differently, and this can cause problems. Sometimes what we say can lead to major communication problems. The receiver may mistakenly believe she knows what the sender means, may own poor listening skills, may not be interested, or may not understand the sender's perceptions.

2. Semantic Barriers

Semantics refers to word choice. The words you choose to get the message across will influence how your message is received. Examples of semantic barriers include poor choice of words, poor sentence structure, incomplete or unorganized thoughts, and inappropriate language.

3. Psychological Barriers

These may include bias and prejudice, preconceived ideas, or strong feelings about a subject held by senders or receivers that lead to distortions in sending and receiving information.

4. Physical Barriers

Noise such as a loud air conditioner, the chattering of voices of people nearby, traffic, and the hum of a computer are all examples that can lead to a communication breakdown.

5. Physiological Barriers

A headache, illness, hunger, or fatigue can lead to difficulties in sending or receiving effective messages.

PRACTICE: IDENTIFYING BARRIERS

Consider this case of miscommunication. After reading this scenario, decide what barriers created this miscommunication.

Constables Stevens and Wu are on patrol duty together. They are working the night shift and have been dispatched to investigate a bar brawl. When they arrive at Sam's Tavern, the officers immediately notice that a large crowd is standing in a circle, cheering and watching two men in a physical altercation. The officers walk toward the crowd shouting that they need to get through. Music is blaring, and no one is moving. The officers eventually make their way through the crowd and break up this fight between two drunken patrons. Stuart Scott is lying on the ground, unable to get up from the floor. The crowd cheers for Bill Blake, who appears to have injured Scott in the fight.

Constable Stevens takes patron Blake aside and asks him to explain what happened. Blake answers, "I sure gave Scott a good left hook in the mouth. I bet I knocked two teeth out. The crowd sure loved that one! Did you see the blood? I'm the best fighter around! Ain't anyone around here that can beat me! What about another beer? Can I get myself another, officer?"

Constable Stevens then repeats his request. This time Blake replies, "What's happenin'? Well, Scott is lying on the floor, I'm talkin' to you, and everyone else in the bar seems to be having a few beers and a good time. That's all, officer." Constable Stevens repeats his request for a third time and waits for a response. Blake answers, "It was the best fight people around here have seen in a long time. It sure was exciting."

At this point, Constable Stevens places Bill Blake under arrest for assault.

List the barriers that created this miscommunication.

OVERCOMING BARRIERS

We sometimes cannot overcome all the barriers that may interfere with the communication process. Let's face it; we do not always have control over noise levels or whether someone speaks our language. However, there are a few things we can do to make us better communicators.

1. Realize that communication is imperfect. Barriers to communication do exist, and they sometimes come into play.

2. Gear your message to your receiver. For example, if you were giving a presentation to a group of third-graders, you would gear your language and presentation differently than if you were giving a presentation to a group of college students.

3. Work on yourself as a communicator. Improve your language and listening skills.

4. Always plan for feedback. If you are the sender of the message, make sure you allow for feedback in the communication process.

ORAL AND WRITTEN COMMUNICATION IN POLICING

Both written and oral communication are vital parts of law enforcement. Think about it. As a police officer, communicating effectively is crucial. Everything a police officer does, in one way or another, involves a great deal of communication, whether oral or written. A police officer uses oral and written communication routinely. We can begin to see the importance of effective communication skills in the law enforcement profession when we begin to look at whom the officer deals with on a regular basis.

First, a police officer uses her communication skills with **victims, witnesses, and suspects.** A police officer uses both oral and written communications with these individuals. The officer uses oral communication skills first. Interviewing individuals, listening attentively, and questioning carefully are all aspects of oral communication that an officer needs to take into consideration. Next, written communication skills are used by the police officer. Recording the critical information needed to complete a written police report is one of the most important tasks a police officer performs.

Lawyers and judges are interested in what a police officer has to say about a particular case or what an officer has written in his notebook and police report. What the officer has to say about a case, whether it is delivered orally while in the box in court or it is written in the officer's notes and report, will have some influence on the outcome of a case. The Crown attorney will be listening or looking for information that may lead to a conviction, and the defending lawyer will be looking for any loopholes in what a police officer says or has written that may lead to the dismissal of the case.

Jurors are also attentive to what a police officer has to say about a particular case. An officer is a witness in court. Jurors listen to the officer's testimony. Again, what an officer has to say in court may have some influence on the jurors and on the outcome of the case.

Media personnel include individuals from radio, television, and newspapers. They will focus on what the police officer has to say about a particular case. The media will want to report news that they deem to be in the public interest. In this situation, oral and written communications are used. The media may interview the officer or may contact police department personnel who handle media situations.

The police department may include your staff sergeant, police chief, or fellow police officers. A police officer's oral and written communication skills come into use. The staff sergeant or police chief may be interested in reading the officer's reports to keep up to date on a particular case or the daily situations a police officer faces. A police officer may be asked to describe orally these same situations to her superiors. Fellow officers may also need to read police reports if a case is being handed over to them. A police officer may also write memos to superiors or co-workers to deal with internal matters, or he may write letters to deal with external matters.

Corrections personnel include prison staff and probation staff. These parties often look at police reports to get an overall sense of what happened in a criminal's case and to determine what needs to be done with the inmate or criminal.

These are just some of the people with whom a police officer may communicate. Both oral and written communication skills are used extensively by a police officer; therefore, it is important that she be able to communicate effectively. An officer must be able to get his message across to the receiver, whether that message is conveyed orally or through writing. In many cases, an officer's ability to communicate effectively will determine whether a criminal is put behind bars or set free. Therefore, a police officer needs to make communication skills a top priority. Remember, no one becomes a good writer or speaker overnight. Like anything else, these skills take a lot of practice.

DISCUSSION QUESTIONS

1. Explain the communication process.
2. What are some of the barriers that can interfere with the communication process?
3. How can one overcome these barriers?
4. Think of a situation that you have encountered in which barriers interfered with the communication process. Explain what happened, what barriers occurred, and how they could have been avoided.
5. Think of yourself as a communicator. How could you improve your communication skills?

ACTIVITY: BARRIERS

Scenario One

Constable McKenzie is working the night shift. While patrolling his given area, he notices a driver who appears to be speeding and swerving between lanes. Constable McKenzie decides to pull the driver over. When the driver steps out of his car, Constable McKenzie notices that he is unsteady on his feet. The driver says, "Clearly, officer, I've had a few to drink." Constable McKenzie then notices that the driver's words are slurred and charges him with impaired driving.

Later that evening, Constable McKenzie dictates his report regarding this situation. In his dictation, Constable McKenzie makes a point of recording exactly what the driver said.

When the case goes to court, a problem occurs. The defence lawyer has obtained and reviewed a copy of the police report. Constable McKenzie is in the box, being questioned by the defence lawyer. With a copy of the police report in hand, the lawyer says, "Constable McKenzie, isn't it true that Mr. Smith was clear in his speech when he spoke to you?" Confused by the question, Constable McKenzie answers, "No, this isn't true. Mr. Smith was slurring his words." The defence lawyer proceeds by saying, "Well, let me refresh your memory, Constable McKenzie. In your police report you wrote, 'Mr. Smith spoke clearly when he said, 'I've had a few to drink.'" Constable McKenzie was horrified. What he had dictated had been misinterpreted. This unfortunate miscommunication led to the dismissal of the case.

QUESTIONS

1. What went wrong here?
2. How could this situation have been avoided?

_____

_____

_____

_____

_____

Scenario Two

Here is another scenario involving miscommunication. After reading the scenario, decide what barriers created this miscommunication.

Constable Chow is dispatched to a break-and-enter call. When he arrives at the scene, he notices that the front window of the house is smashed in and the door is banged in. Upon entering the house, he notices that the house is ransacked. He also notices Mrs. Alfonso, owner of the house, sitting in a corner crying. Mrs. Alfonso appears to be extremely upset about the break-in. Constable Chow approaches Mrs. Alfonso and asks, "Mrs. Alfonso, are you alright?" Mrs. Alfonso does not respond and continues to cry. Constable Chow continues, "Mrs. Alfonso, do you know what happened here?" Again, Mrs. Alfonso does not answer. She continues to cry and to stare straight ahead. At this point, Maria Scalli, daughter of Mrs. Alfonso, enters. She approaches Constable Chow and tells him that her mother does not speak English. She also tells Constable Chow that her mother is manic-depressive. Constable Chow calls the station for assistance.

What barriers interfered with the communication process?

_____

_____

_____

_____

_____

What can Constable Chow do to aid the communication process?

_____

_____

_____

_____

_____

Scenario Three

You are a constable working a night shift, and you have been dispatched at 1130 hours to Louie's Pizza Parlour to take a report on a customer disturbance. When you arrive at the scene, you find a patron who appears to be intoxicated sitting at a table. He is slurring his speech and is unable to stand up. He is also holding a bottle of wine from which he is drinking. The owner of the pizza place, Louie Dovigi, tells you that the customer came into the pizza place over two hours ago and has refused to leave. The man continually asks for food but has no money. When you approach the man, you ask him his name and address. The man does not answer your questions, but instead asks you for five dollars so he can buy some pizza. You ask him the questions again and the man does not answer. He continues to ask for a piece of pizza. At this point, you call for assistance and take the man to a homeless shelter.

What barriers interfered with the communication process?

_____

_____

_____

_____

_____

Scenario Four

You are a constable and you have just finished working a 12-hour night shift. You are about to go home to get some sleep when your staff sergeant calls you in for a brief meeting. He had reviewed a report you had submitted and he asks you to make some corrections and resubmit the report as soon as possible. You take the report and go home to get some sleep. You return to work on your next scheduled work day and go to your staff sergeant's office to resubmit the report. Your staff sergeant is angry. He wanted the report resubmitted the day he had returned it to you.

What barriers interfered with the communication process?

_____

_____

_____

_____

_____

How could this situation have been avoided?

_____

_____

_____

_____

_____

ACTIVITY: USING COMMUNICATION SKILLS IN LAW ENFORCEMENT

List the various scenarios or situations a police officer encounters on a daily basis on his routine patrol.

_____

_____

_____

_____

Now, review your list, and write down how communication skills are used for each scenario.

_____

_____

_____

_____

As a future police officer, how do you plan on improving your communication skills?

_____

_____

_____

_____

Chapter 2

Memo Writing

LEARNING OUTCOMES

Upon successful completion of this chapter, you should be able to

1. define the purpose of the memo
2. recognize and correctly use the memo structure
3. recognize and correctly write the various police type memos
4. evaluate the effectiveness of the memo produced
5. use the writing stages to produce a memo
6. edit and revise the content of the memo
7. recognize and correctly use the parts of the memo
8. recognize and correctly use the appropriate memo format
9. recognize and correctly use the structure of the body of the memo
10. recognize and apply the guidelines to memo writing
11. recognize the advantages and disadvantages of using email
12. correctly write an email

PURPOSE OF THE MEMO

In any organization, whether a police department, financial institution, or volunteer agency, co-workers need at times to communicate in writing. There may be situations when a simple phone call or direct contact is not enough. You may need to have information recorded in writing and filed away for future reference. Or sometimes it is not possible to see someone face to face to get your message across. This is when the memorandum (memo) comes into use. The memo is used within an organization to convey a message to an individual or a group of people. It keeps information flowing within an organization. You may need to ask your supervisor for some time off, to announce an upcoming meeting, or to explain a particular situation. These matters are communicated through the use of the memo. A memo is used internally. You do not send a memo to someone outside your organization.

PARTS OF THE MEMO

The memo has standard headings that are found at the top of the page. These headings should be the first thing your reader sees. Your organization may have a standard memo on company letterhead that simply needs to be filled in. In other organizations, you may need to complete the memo headings on your own.

There are four standard memo headings. They are

To:
From:
Re:
Date:

Look closely at what each heading contains.

To: To whom are you writing? Remember you want to remain professional. Even though you may know the person you are writing to personally, keep in mind that you are communicating on a professional level. Therefore, address the person by his or her full name. Also, include the person's professional title. Here are some examples:

To: Constable Nick Ugami
To: Sergeant Ramona Streeter

To: Ellis Jones, Accounting Clerk
To: Constable Sylvia Bouchard
    Constable Blaine Black
    Constable Lyle Macintyre
    Constable Daniel Chan

From: You place your name, as the sender of the message, here. Again, keep in mind that you are approaching this message on a professional level. Use your full name and title. After your memo has been typed, place your initials beside your name. Do not sign your name anywhere on the memo. Placing your initials beside your name will verify to the reader that you wrote the memo. Here are some examples:

From: Constable Sally Armstrong S.A.
From: Sergeant Lucas Niprut L.N.
From: Chief Nester J.N.

Re: This refers to the subject or reference. What is the subject of your memo? For this heading, you need to think of a shortened version of your subject. The reader of the memo, in most cases, will want to know what the memo is about before he or she reads it. The "Re" line is the place where the reader will look for this information. Make sure the subject you choose clearly reflects your main message. Also, remember to keep it short. This should be a quick reference for your reader. Some examples of the reference line include

Re: Health and Safety Committee Meeting
Re: Upcoming Self-Defence Workshop
Re: Vacation Schedule

Date: This is the final heading seen at the top of the memo. The date you are writing the memo is placed here. You may write the date in full or use a notation system. Go with the style that is used and accepted by your organization. Here are some examples:

Date: July 7, 2005
Date: 02 05 05
Date: Friday, May 15, 2005

BODY OF THE MEMO

The body of your memo is where your message goes. What is it that you need to tell your reader? The body of your message may be organized in the following way:

First Paragraph

Your introductory paragraph is what will be read first, so put your main message here. This is not creative writing. Get to the main point quickly. When your reader picks up your memo and starts to read, he or she will want to know right away what it is that you want. So do that in your first paragraph. Give your reader the most important message first.

Middle/Body Paragraph(s)

Once you have given the reader the main message, you will then have to give her details and explanations. These may include dates, locations, what people need to bring, and who is involved. To help you get your details and explanations together for this paragraph, remember your five Ws. Answer the questions *who, what, when, where,* and *why*. It may take you only one body paragraph to give your details and explanations, or it may take you more, depending on what your subject is and what you need to get across to your reader.

Closing Paragraph

In the closing paragraph, you need to do two things. First, tell the reader what it is you want him or her to do. Second, before ending, look to the future and end in a positive way. Don't forget to give your reader the telephone extension number where you can be reached.

SKETCH OF THE MEMO

Here is what a sketch of a memo looks like:

To:
From:
Re:
Date:

> ### Introductory Paragraph
> Give your main message.

> ### Body Paragraph
> Give the details and explanations.

> ### Concluding Paragraph
> Tell your reader what he needs to do.
> End in a positive way.
> Offer your reader an extension number where you can be reached.

PRACTICE: MEMO PARAGRAPH STRUCTURE

List the information that should be included in each of the paragraphs of a memo for the following scenarios. Create your own details for each scenario.

1. Announcing a health-and-safety meeting to a group of officers

_____

_____

_____

_____

_____

2. Asking your staff sergeant for vacation time

_____

_____

_____

_____

_____

3. Informing constables of an upcoming shift change (shifts will revert to eight-hour stints instead of twelve-hour ones)

_____

_____

_____

_____

_____

4. Informing your sergeant about your concerns regarding a domestic dispute

_____

_____

_____

_____

_____

5. Making a request of your sergeant that repairs be done to your cruiser

_____

_____

_____

_____

_____

MEMO FORMAT ONE

Look over the following memo to get an idea of how a memo is set up. Follow this format when you are using plain paper. No letterhead appears at the top of the page.

To: Constable Liam McGillvary
From: Sergeant Elaine Miller E.M.
Re: Success of Child Find Program
Date: February 13, 2006

I would like to commend you on your efforts in organizing the Child Find Program this year. Your hard work, enthusiasm, and commitment to the program were evident in its huge success this year.

Over 2000 children registered for the program, and based on the comments received from parents and workers at the event, all went smoothly and efficiently. I have received your memo outlining some of the changes you would like to implement next year, and they sound as if they would work wonderfully!

Again, thanks for a job well done. I look forward to working with you to implement some of the changes discussed for next year's event. I'm sure it will be an even bigger success. Please contact me at extension 8776 to discuss these changes further.

MEMO FORMAT TWO

Look over the following memo to get an idea of how a memo is set up. Follow this format when you are using letterhead.

Metropolitan Police Services

Memorandum

To: Constable Liam McGillvary
From: Sergeant Elaine Miller E.M.
Re: Success of Child Find Program
Date: February 13, 2006

I would like to commend you on your efforts in organizing the Child Find Program this year. Your hard work, enthusiasm, and commitment to the program were evident in its huge success this year.

Over 2000 children registered for the program, and based on the comments received from parents and workers at the event, all went smoothly and efficiently. I have received your memo outlining some of the changes you would like to implement next year, and they sound as if they would work wonderfully!

Again, thanks for a job well done. I look forward to working with you to implement some of the changes discussed for next year's event. I'm sure it will be an even bigger success. Please contact me at extension 8776 to discuss these changes further.

TONE

When you are writing memos, remember that positive language is accepted more readily and favourably. Focus on the use of positive rather than negative words and language. Positive language is uplifting and pleasant. Negative words and language tend to put the reader on the defensive. Positive language focuses on what can be done and what is possible, whereas negative language focuses on what can't be done and what isn't possible. Always analyze what you have to say and then present it in positive language. Your memo will be received more favourably.

FOCUS ON *YOU*

As the writer, you need to put yourself in the reader's position. Every reader wants to know what's in it for him or her. When you are writing, ask yourself, "What's in it for my reader? How will my message benefit my reader?" Once you have answered these questions, write your message, emphasizing the benefits to your reader. Try to use the second person pronouns (*you, your*) in your writing. Using these pronouns emphasizes the benefits your reader will gain.

PRACTICE: TONE AND *YOU* FOCUS

Read the following memo and make changes so that the tone is positive and a *you* focus is apparent.

Metropolitan Police Department
Memorandum

To: Constable Garson
From: Staff Sergeant Pollie s.P.
Re: Police Reports
Date: April 18, 2006

I am writing this memo to let you know that there have been some problems that have occurred in the way you write your police reports. As far as I am concerned, the problems occur because of your carelessness.

Every time I receive a report that was written by you, I can count on the report being written in a sloppy way. I always note that there are problems with your spelling and grammar. Most importantly, I always know that the report will be written in a subjective manner. I want you to do something to clean up your reports. I suggest that you take a grammar clean-up course and a report writing workshop.

I want you to take care of this as soon as possible. Submit to me a schedule you plan on following to rectify this problem. I can be reached at extension 870.

STAGES IN WRITING THE MEMO

In any type of writing, you need to go through three stages. These are

- Prewriting
- Writing
- Rewriting

Prewriting

In this stage, you gather all the points you will need to include in your memo. Remember that your memo needs to be complete, so think of everything you need to tell the reader. Make a list of all your points. When you are making your list, you don't need to have your points in the right order. You are just jotting down all your ideas. Once you have listed all your points, review your list. Cross out any ideas that do not need to be in your memo. Check off the points you will need. This will give you the opportunity to review your list to ensure that all your points are there.

The next step in the prewriting stage involves filling in your outline. Using your outline will help ensure that all the necessary information is included in the memo and that it is in the right place. Take steps to avoid producing an incomplete memo. You would not want to have to send a follow-up memo saying, "By the way, I forgot to mention in my last memo that . . . " Your writing is a reflection of you. People are going to make judgments about you based on your communication skills. Therefore, you want to send out a memo that is complete. Here is an example of an outline you may want to use when you are doing your prewriting for the memo. Using a format such as this one will ensure that your memo says everything you need it to say.

MEMO WRITING OUTLINE

Introductory Paragraph

Important Message: _____

_____

_____

Body Paragraphs

Details and explanations: (*Remember who, what, when, where, why.*) _____

_____

_____

Concluding Paragraph

What do you want the reader to do?_____

_____

_____

Look to the future/end positively. _____

_____

_____

Offer a contact extension number. _____

_____

_____

Writing

Once you have completed your outline, you can move into the writing stage. With your outline as a guide, begin writing. This is your first draft, so it doesn't need to be perfect. At this point, your goal is to get your ideas down in written form.

Rewriting

Once the rough draft is complete, you can begin looking at it for revisions. You need to evaluate what you have written. Ask yourself a series of questions:

- Does the memo contain all the appropriate headings? Are they complete?
- Is the format correct?
- Does the introductory paragraph contain the main idea?
- Do the body paragraphs contain the details and explanations?
- Does the concluding paragraph tell the reader what action needs to be taken? Does it end positively, look to the future, and offer a telephone extension number?
- Is the memo complete and accurate?
- Are there any grammatical problems that need to be fixed?

While you are evaluating your memo, make any necessary changes on your rough draft. These changes will lead to your next copy. Each copy you write will be a revised draft. Evaluate each draft until you are satisfied with a copy that will become final.

STEPS IN THE MEMO-WRITING PROCESS

Read the following scenario and follow through to the finished product to see how this police officer wrote her memo.

Sergeant Madeline Pavoni has finalized some training plans for the constables at the detachment. The training consists of a report-writing workshop. Many officers have approached her and expressed interest in participating in a workshop to improve their report-writing skills. Sergeant Pavoni has managed to contact a report-writing consultant to come to the detachment to do the training. The consultant, Beverly Barsanti, has presented this workshop to many other police services across the province and has received favourable responses. Ms. Barsanti has agreed to conduct four workshops, lasting three hours each, for the detachment during the week of May 8, 2006. Sergeant Pavoni now needs to relay this information to the constables in her detachment. Here is the process Sergeant Pavoni followed to complete the memo she sent to the constables:

1. Making a List

- many officers have requested some training in report writing
- a report-writing workshop will take place the week of May 8–11, 2006. Four workshops, each lasting three hours, will be offered
- interested officers should fill in the attached registration form and submit it to the training department
- officers will be given time in lieu for participating in the workshop
- presenter will be Beverly Barsanti, a consultant in report writing

- she has presented this workshop to many police services across the province and has had success
- workshop will take place in Training Room 145
- training will be available to the first 50 officers who register
- facilitator will arrive in the city on May 7, 2006
- facilitator is asking for $1000.00 to conduct the workshop
- juice and muffins will be provided as a snack
- workshop will take place from 1300 until 1600, Monday through Thursday
- facilitator has great expertise in the area of report writing
- workshop will benefit officers — hands-on approach
- participants will receive practical writing experience and feedback from the facilitator
- call extension 9630 for questions or comments

Sergeant Pavoni's next step involved going over her list and checking off the points she felt needed to be included and crossing out the points that weren't necessary. Look at what she did with her list:

- many officers have requested some training in report writing ✓
- a report-writing workshop will take place the week of May 8–11, 2006. Four workshops, each lasting three hours, will be offered ✓
- interested officers should fill in the attached registration form and submit it to the training department ✓
- officers will be given time in lieu for participating in the workshop ✓
- presenter will be Beverly Barsanti, a consultant in report writing ✓
- she has presented this workshop to many police departments across the province and has had success ✓
- workshop will take place in the Training Room 145 ✓
- training will be available to the first 50 officers who register ✗
- facilitator will arrive in the city on May 7, 2006 ✗
- facilitator is asking for $1000.00 to conduct the workshop ✗
- juice and muffins will be provided as a snack ✗
- workshop will take place from 1300 until 1600, Monday through Thursday ✓
- facilitator has great expertise in the area of report writing ✓
- workshop will benefit officers — hands-on approach ✗
- participants will receive practical writing experience and feedback from the facilitator ✓
- call extension 9630 for questions or comments ✓

2. Using an Outline

Once Sergeant Pavoni chose the ideas that needed to be included in the memo, she then made an outline before beginning to write. Using an outline helped Pavoni to see whether there were any ideas she had missed. Here is what her outline looked like:

SAMPLE MEMO-WRITING OUTLINE

Introductory Paragraph

Important Message - many officers have requested training

- report writing workshop will take place

Body Paragraph(s)

Details and explanation: (remember who, what, when, where, why)

- training will take place week of May 8–11
- four, 3-hour sessions will occur
- daily from 1300-1600
- Room 145

- officers participating will be given time in lieu

- presenter is Beverly Barsanti, report-writing consultant

- extensive report-writing background

- presented to numerous police services

- workshop will provide practical writing experience and feedback

Concluding Paragraph

What do you want the reader to do: officers fill in registration form and submit it to training department

Look to the future/end positively: look forward to seeing you there

Offer a contact extension number: call extension 9630 — questions/ comments

3. First Draft

With the completed outline by her side, Sergeant Pavoni began writing her first draft. This is what she came up with:

Metropolitan Police Department
Memorandum

To: All constables
From: Sergeant Pavoni
Re: Report Writing Workshop
Date: February 10, 2006

The training department of the Metropolitan Police Department is sponsoring a report writing workshop. There have been many requests in the passed for some raining in this area.

The workshop will take place from Monday, May 8, 2006, to Thursday, May 11, 2006, from 1300 to 1600 daily. The training will take place in room 145. Officers who participate in the training will be given time in lieu. The facilitator of the workshop will be Beverly Barsanti, a consultant in report writing. She has an extensive report writing background and has presented many workshops to police services across the province. The workshop will provide practical writing experience and feedback from the facilitator.

Officers who are interested in taking part in this workshop should fill in the attached registration form and submit it to the training department. I look forward to seeing you there. If you have any questions or concerns, feel free to contact me at extension 9630.

---

4. Editing

With the first draft complete, Sergeant Pavoni began the rewriting process. She read her memo and made the necessary changes. Here are the changes she made to her first draft:

Metropolitan Police Department
Memorandum

To: All constables
From: Sergeant Pavoni
Re: Report-Writing Workshop
Date: February 10, 2006

not necessary – memo is internal

The training department ~~of the Metropolitan Police Department~~ is sponsoring a report-writing workshop. There have been many requests in the passed for some training in this area.

re-word

past

The workshop will take place from Monday, May 8, 2006, to Thursday, May 11, 2006, from 1300 to 1600 daily. The training will take place in Room 145. Officers who participate in the training will be given time in lieu. The facilitator of the workshop will be Beverly Barsanti, who is a consultant in report writing. She has an extensive report-writing background and has presented many workshops to police services across the province. The workshop will provide practical writing experience and feedback from the facilitator.

when does this need to be submitted

Officers who are interested in taking part in this workshop should fill in the attached registration form and submit it to the training department. I look forward to seeing you there. If you have any questions or concerns, feel free to contact me at extension 9630.

---

5. Final Copy

Once Sergeant Pavoni edited her first draft, she kept writing and editing until she came up with a copy she was satisfied with. Here is the final copy of the memo:

Metropolitan Police Department

Memorandum

To: All constables
From: Sergeant Pavoni   M.P.
Re: Report-Writing Workshop
Date: February 10, 2006

The training department is sponsoring a report-writing workshop for interested officers. In the past, many officers have requested training in this area.

The workshop will take place from Monday, May 8, 2006, to Thursday, May 11, 2006, from 1300 to 1600 daily. The training will take place in Room 145. Officers who participate in the training will be given time in lieu. The facilitator of the workshop will be Beverly Barsanti, a consultant in report writing. She has an extensive report writing background and has presented many workshops to police services across the province. The workshop will provide practical writing experience and feedback from the facilitator.

Officers who are interested in taking part in this workshop should fill in the attached registration form and submit it to the training department before April 28, 2006. The workshop will greatly benefit those attending. If you have any questions or comments, please call extension 9630. I look forward to seeing you there.

PRACTICE: COMPLETING MEMO A

Sergeant Morin is in the process of writing a memo to the constables in her division regarding the formation of a new committee in the department. Sergeant Morin needs to recruit two constables within her division to volunteer to become members of a committee looking at ways to resolve the downtown loitering problem that has been going on for some time. People loiter on the downtown streets until all hours. This situation has caused problems for tourists. Sergeant Morin got as far as listing possible ideas that she could include in her memo. Review her list of ideas and choose those you think should go into the memo. You may also add ideas that you feel are necessary to complete the memo. Then finish the rest of the writing process and complete the memo that Sergeant Morin might send out.

Sergeant Morin's Ideas

- looking for constables who are willing to become members of a committee that will look at the problem of loitering in the downtown area
- loitering has been causing problems for the tourists
- committee will meet once a month for two hours
- community image is suffering
- police officers need to become more visible in the community
- the incident that happened last month with Constable Meghan does not help us as a department
- being a member of this committee would be a good thing to include on your resumé if you wish to become a sergeant
- meetings will take place in the boardroom
- meetings will take place the first Monday of each month
- the committee will review the current situation and make recommendations to Chief Edwards
- Chief Edwards will then review the recommendations and make decisions about implementation
- part of the responsibility of the committee will be to include a schedule for implementation
- dinner may be provided for the meetings
- I will not be able to participate in the committee because of other commitments

PRACTICE: COMPLETING MEMO B

Below is the introductory paragraph of a memo. Complete the memo by adding the middle paragraph and the closing paragraph.

Metropolitan Police Department

Memorandum

To: Staff Sergeant Bellows
From: Constable Miggins   A.M.
Re: Shift Change
Date: January 9, 2006

I would like to request a shift change the week of March 13 to March 17, 2006.

_____

_____

_____

_____

_____

_____

GUIDELINES FOR MEMO WRITING

Keep these guidelines in mind when writing a memo:

- **Be Brief** — Give the main message and move on. The reader doesn't want to know more than he or she needs to know.
- **Be Complete** — Make sure that all necessary information is included in your memo. You don't want someone calling to say that you missed something; your sending a follow-up memo can look bad.
- **Be Clear** — The intent of the memo and its supporting statements should be evident to the reader.
- **Be Professional** — Make your memo look presentable, and type it. Poor appearance may portray a careless writer.
- **Be Correct** — Proofreading for grammatical errors and poor sentence construction is important. A memo with a spelling mistake may portray a careless writer.

USING COMPUTER TEMPLATES TO WRITE MEMOS

Many computer programs come equipped with memo templates. Check out the templates your computer program offers; you may want to use these if they provide what you need.

USING GRAMMAR AND SPELLING CHECKERS

Computer programs come equipped with grammar and spelling checkers. The grammar and spelling checkers are a good resource to use. However, do not rely on them completely. Here are some tips:

- Run through grammar and spelling checkers twice.
- Print your documents and review it from the hard copy. You are more likely to catch an error when you are looking at the hard copy than the screen.
- Keep in mind that sometimes the grammar and spelling checkers will recognize something as an error that really isn't an error. Always consult reference books, dictionaries, or ask someone to help you edit your document if you are unsure.

TYPES OF POLICE MEMOS

As a police officer, you may come across certain situations that require you to write a memo. The memos you write can be classified into three different types. These are

- informational memos
- announcement memos
- request memos

Information Memo

This memo is written in situations requiring that information be distributed to an individual or a group. An informational memo may be written to report organizational policies and procedures, to relay departmental news or updates, to make suggestions/recommendations,

Metropolitan Police Department
Memorandum

To: All constables
From: Chief Ben Buckley B.B.
Re: Bullet-proof Vest Policy
Date: April 7, 2006

As of May 1, 2006, new departmental policy will require officers to wear the new bullet-proof vest at all times while on duty.

This new policy has been put into effect to ensure that all officers are working under the safest conditions possible. Other police departments around the province have been practising this policy for some time now and are pleased with the results. We require that you get sized for your vest on April 14, 2006, in Board Room 102, between 0800 and 1700.

Once again, as of May 1, 2006, this new policy will come into effect. If you have any questions or comments regarding this new policy, feel free to contact me at extension 453.

Metropolitan Police Department

Memorandum

To: Constable Lawrence Mah
      Constable Vivian Bouchard
      Sergeant Miki Mullin

From: Sergeant Yolanda Pickering   Y.P.

Re: Health and Safety Committee Meeting

Date: June 5, 2006

A health and safety committee meeting will be held on Monday, June 19, 2006, from 1400 to 1700 in Room 502.

The meeting has been scheduled to address new funding that is available to improve health and safety in an organization. In order to be eligible for the funding, we must submit a proposal outlining initiatives we would like to take to address health and safety concerns. This is an excellent opportunity for us to make some of the changes we have been talking about. Please come to the meeting prepared to work on a proposal.

Please confirm your attendance at the meeting by phoning me at extension 435. I hope to see you there. I look forward to working with you.

Metropolitan Police Department

Memorandum

To: All constables
From: Sergeant Ted Haskell   T.H.
Re: Workers Needed for Community Dance
Date: January 13, 2006

The Big Bear Community Centre will be holding its annual Community Day Family Dance as part of the winter carnival, and its organizers have once again requested that officers police the event.

This year the Community Day Family Dance will be held on Friday, February 17, 2006, from 2000 to 2300. Officers have been requested to patrol the dance and prevent any problems from occurring. Officers working at this event may either ask for overtime pay or time in lieu.

Anyone interested in working at this event is asked to contact me at extension 890. Your cooperation in this matter is greatly appreciated.

or to make known good or bad news. This type of memo usually dos **not** require a response from the reader.

Announcement Memo

This memo is written to announce upcoming meetings, workshops, or conferences. It also may be written to announce an event that the police department is getting involved in. Such a memo usually does require a response from the reader.

Request Memo

The writer of this type of memo may need time off, a shift change, or permission to attend a conference or workshop. The staff sergeant also may use this type of memo to recruit volunteers for a community event or to recruit officers who are willing to take on some overtime shifts. This type of memo does require a response from the reader.

PRACTICE: WRITING POLICE MEMOS

Use the three writing stages to prepare a memo for each of the following scenarios:

Information Memo

A. The baseball game is now over, and your team beat the local firefighters. Write a memo to the police department staff informing them of the results of the game. Remember to include particulars about how much money was raised for the cancer society and who won the game. You also will want to thank the players and volunteers for taking part in this year's event.

B. Constable Peter Swanson has been promoted to sergeant. Write a memo to the policing staff informing them of this change.

Announcement Memo

A. You now have your baseball team in place. You have been practising for weeks, and the event is approaching. Write a memo to the staff of the police department announcing that the baseball game will take place. In your memo, give your readers the particulars about the event. When preparing this memo, remember your 5 Ws.

B. The annual police conference is taking place soon. As the staff sergeant of your unit, it is your responsibility to inform your officers of the upcoming conference. Write your officers a memo announcing the conference and informing them that they are welcome to submit a registration form to attend this annual conference.

Request Memo

A. You are a sergeant working at the Metropolitan Police Department. Annually, you organize a baseball game against the local firefighters to raise money for the cancer society. As part of this year's plans, you need to begin organizing a team for the annual event. Write a memo to the constables, sergeants, and staff sergeants requesting they volunteer to be part of the baseball team.

B. You are a staff sergeant working at the Metropolitan Police Department. The annual bike registration is taking place next month, and you need officers to volunteer to help with this event. Write the constables in your unit a memo requesting volunteers to help with this annual event. Don't forget to outline the benefits of volunteering.

ACTIVITY: EVALUATING MEMOS

Evaluate the following memos. Correct any errors that may occur in format, paragraph structure, grammar, tone, or wording.

(A)

Metropolitan Police Department
Memorandum

To: Sergeant Roger Graven
From: Constable Ben Blanchard B.B.
Re: Approval for requested promotion from you

I am requesting this approval for promotion do to the fact that I have been with the force for 19 years. This alone should be reason enough. I have an outstanding arrest record, and have assisted in many undercover operations (drug related). Threw the years, I have also taken many supervisory courses to strengthen my abilities for this position.

In addition to these many attributes, I am well liked by the staff and fellow officers. This quality leads me to believe that in a higher rank, I can assist the detachment with it's ongoing crime prevention. My rapport allows me to interact well with the community at large, as well as clients involved in the detachment.

Let me know of the decision rendered regarding approval for promotion. No doubt—I am the best and should be promoted. If there is any information you need, you can contact me during my shift or at my place of residence (555-9076).

(B)

Metropolitan Police Department
Memorandum

To: Constable John Henry

From: Chief Frances Maniacco

Re: Complaint of Using Excessive Force More than Once

Date:

It was brought to my attention that you have been exerting unnecessary force with regards to recent arrests.

As a representative of this police force, it should be known by you that these actions will not be tolerated. An investigation will be held to verify these acusations against you. Please contact me as soon as possible at extension 342 to set up an appointment with me.

(C)

Metropolitan Police Department
Memorandum

To: Sergeant Mike Davis

From: Constable David Parker

Re: Weapons Training Course that I would like to attend

Date: January 30, 2006

Please give me approval to attend the weapons training course that will be held from June 12–18, 2006, in Toronto. This course is being offered by the RCMP. This course will improve my skills when handling a variety of weapons.

Sign the attached registration form and return it to me before May 1, 2006. If you have any questions, let me know.

(D)

Metropolitan Police Department
Memorandum

To: Chief Reg Gallagher

From: Constable Pauline Booker

Date: February 8, 2006

Please accept my apology for my absence from the mandatory weapons training on February 4, 2006. Due to family obligations I was unable to attend.

DISCUSSION QUESTIONS

1. What is the purpose of the memo?
2. What are the three main parts of the memo, and what belongs in each part?
3. Briefly discuss the steps one should take in order to produce a well-written memo.
4. What are the main types of memos written by police officers? Briefly explain each one.
5. What are the essential parts of a memo?

ACTIVITY: WRITING POLICE MEMOS

Write a memo for each one of the following scenarios:

Information Memo

A. You are a sergeant working at the Metropolitan Police Department. You have just received word from Chief Bernard Stevens that all officers are to keep their notebooks current. There have been problems recently with officers not filling in notebooks right after an incident occurs. This situation has resulted in many offenders not being convicted for lack of evidence and officers' recall. Write the constables of your unit a memo expressing your concerns regarding this situation and informing them of the chief's thoughts on this matter.

B. Chief Felban has noticed that some constables are not keeping their cruisers clean. He has asked you, as staff sergeant, to inform your constables that it is their responsibility to keep their cars clean. Write your constables a memo informing them of the chief's concerns and comments.

Announcement Memo

A. You are a constable working at the Metropolitan Police Department. As part of your duties, you have taken on the role of chair for the Weapons-Training Committee. The committee meets bimonthly and discusses any problems or concerns of officers regarding the weapons they are using and any possible training that should occur. It is time for the committee to meet again. Write the members of the committee a memo announcing the next meeting.

B. You are a staff sergeant working at the Metropolitan Police Department. One of your sergeants is retiring and a farewell social is being planned. Write the constables and sergeants of your unit a memo announcing the farewell social being held for departing Sergeant Wollin.

Request Memo

A. You are a constable working at the Metropolitan Police Department. You have been using the equipment in the station workout room for more than five years and you believe it is outdated. You feel that the department should put some money into getting the equipment updated and that this would be of benefit to the officers of the force. Write a memo to your staff sergeant making this request.

B. You are a constable working at the Metropolitan Police Department. You have noticed that the first-aid kits in the cruisers have not been checked and replaced in some time. Write Staff Sergeant Wong a memo requesting that the first-aid kits be examined and replaced if necessary.

ELECTRONIC MAIL (EMAIL) SYSTEMS

Today, many people use email to get a message to someone within the organization or to people outside the organization. There are advantages to using email:

- it provides a written record and allows the recipient to reread the message before responding
- it permits a speed of transmission that puts regular mail to shame. You can send a message in a matter of seconds and your recipient can respond instantaneously if need be

There are also disadvantages associated with using email:

- not everyone has access to email
- not everyone has a positive attitude about using email
- email can be difficult for some people to use
- email is subject to malfunctions of technology

When using email, adhere to the same format as for the memo. Remember, you want your message to be effective and accurate. The email template provides you with the standard memo headings. All you need to do is fill in the heading fields.

Guidelines for Using Email

- Complete the heading fields at the top of the email.
- Keep a positive tone and provide a *you* focus.
- Keep your email clear, complete, and accurate. You do not want to send a message that is unclear to the reader or that has erroneous information.
- Keep your message concise. Don't give the reader more information than is necessary.
- Keep your email professional. Email messages can easily be forwarded or copied to other readers, so be careful with the kinds of messages you send.
- Proofread your email before you send it. Check for spelling and grammatical errors.

When replying to a multiple-reader message, ensure that you want everyone listed to receive your response. It may be easier sometimes to click the *Reply All* button, but before doing so, make sure that you want everyone listed to get your response. Take the time to address your reply to only those readers who need it.

- Keep your email address a professional one if you are writing on behalf of your organization. Some addresses would be considered inappropriate and looked upon negatively. As an example, using an email address such as copperdude@metropolitanps.ca would likely not be received favourably.
- There are some slang, short forms, and vocabulary that many emailers are using that would be viewed as unprofessional. Some examples include using *i* for *I* or *u* for *you*. Also, changing the ending of *–ing* words to *–in* and changing words with the letter *s* to *z* would not be seen as professional. Some examples include using *walkin* instead of *walking* and *waz* instead of *was*. Finally, avoid using acronyms.

Some examples include *lol*, which has become known as the short form for laugh out loud; and *zup* meaning *what's up*. Again, using these expressions in professional writing would not be received in a positive way.

- Keep a professional tone when writing your email. You want the recipient of the email to think of you as a professional. Maintaining a professional tone will give you credibility.

- Use a three-paragraph structure to ensure that the message is complete. See the format shown below.

Here is a summary of what the email should include:

To: Receiver's email address
From: Sender's email address
Subject: Main topic of the email

Body of the Email
Opening Paragraph
– Gives the reader the main idea/message behind the email

Body Paragraphs
– Gives the reader the details of the email
– Is logical, clear and coherent to the reader

Closing Paragraph
– Tells the reader what he needs to do
– Ends in a positive way

Attachments
– Make reference in your email to your reader that there is an attachment

---

PRACTICE

A. Constable Fellinger wrote and sent this email to Staff Sergeant Bourdage. Constable Fellinger is asking the staff sergeant for permission to attend a provincial officers workshop on arrest procedures. Read the email and list the many errors you see.

To: sbourdage@metropolitanps.ca

From: topcopfellinger@metropolitanps.ca

Re:

Hi Sarg–

Zup? i want to know if u would let me take some time off to attend the arrest procedurez workshop in t.o. It will B useful to me since u have told me that lately ive bin doin the wrong thing when i make an arrest. Not all the time though – lol. The workshop is 2 dayz long. The guyz puttin it on say its worth the time. Like i have already said, its in t.o. September 5 to September 7. $200 bucks. U can pay for the workshop fee and hotel and i will drive my car and pay for gas. Compromize? U will not be sorry if u give me permission to go. Sound good? Let me no. Thatz it.

What errors did Constable Fellinger make?

_____

_____

_____

_____

_____

Rewrite the email, correcting the errors.

_____

_____

_____

_____

_____

_____

_____

_____

_____

_____

_____

_____

_____

_____

B. You are a staff sergeant working at the Metropolitan Police Department. Send an email to all staff sergeants of the force informing them that a special meeting is being called to discuss changes in their retirement packages.

C. You are a staff sergeant working at the Metropolitan Police Department. Along with Staff Sergeant Mabri, you have taken the responsibility for making changes to the station's policies and procedures. Send Sergeant Mabri an email asking her when a convenient time would be for the two of you to meet on a weekly basis.

ACTIVITY

A. You are Chief Donovan working at the Metropolitan Police Department. You have noticed that your staff sergeants have worked extremely hard lately getting the station ready for an upcoming conference. Send your staff sergeants an email inviting them to an appreciation luncheon.

B. You are Chief Stern working at the Metropolitan Police Department. One of the constables in your department has been off from work due to a hand injury. He has been away for some time, and you had thought he would have returned to work by now. Write Staff Sergeant Mobilio an email requesting an update on this individual's condition and situation.

C. You are a staff sergeant working at the Metropolitan Police Department. You have noticed that some of your constables are not getting their police reports in on time. You think they need a friendly reminder about the department's policy on submitting reports. Write a general email to your constables reminding them of the department's policy on the submission of police reports.

D. You are a constable working at the Metropolitan Police Department and are interested in putting together a baseball team of police officers to participate in the annual cancer fundraiser. Send an email to the police officers inviting anyone who may be interested in participating to attend a meeting.

Letter Writing

Upon successful completion of this chapter, you should be able to

1. define the purpose of the letter
2. define and properly use the parts of the letter
3. recognize the sketch of a letter
4. recognize and use the various letter styles
5. recognize and use open punctuation and two-point punctuation with letters
6. properly structure the body of the letter
7. recognize and use the letter format
8. use the prewriting, writing, and rewriting stages of the letter process
9. recognize the different types of police letters
10. write different types of police letters
11. effectively use the computer to write letters

PURPOSE OF THE LETTER

Letters are written communications that are sent externally. If you have a message to relay to someone outside your organization, you would send her a letter. Written communication in the form of letters is essential to provide a permanent communication record.

PARTS OF THE LETTER

There are eight major parts to a letter. These are the return address or letterhead, the date, the inside address, the salutation, subject lines, the body, the complimentary closing, and the identification line and signature. Other parts that may be included in the letter are the attention line, the typist's initials, and the enclosure.

Return Address

This is your address as the writer of the letter. If you are using company letterhead, then the return address will already exist. The letterhead will have the company's name and address on it. If you are sending a personal letter, you will type your full address. Do not include your name with the return address.

Date

The date you are writing the letter appears here. You may write the date out in full (February 27, 2006) or use the notation system (2006 02 27). Use the system your organization uses and accepts.

Inside Address

The inside address includes the name of the person to whom the letter is being sent (addressee), that person's business title, the name of her company, and the full address, including the postal code. Each item should be on a separate line, unless the person's name and title are short enough to fit on one line without extending it too far.

Salutation

The salutation is the opening to the letter. "Dear" is still the most common and accepted opening, although there is a trend that drops the "Dear" and uses the name only, such as "Ms. Lee." Only when you are sending a personal letter and know the addressee personally should you use the first name. If you are sending a letter on behalf of your organization, you want to keep it professional. In most cases, you should address the person by her last name. It is important to try to verify the name of the person you are writing and, if your addressee is female, whether she prefers Ms. or Mrs. If you are writing to a department, you may want to use Dear Sir(s), Dear Sir and Madam, or Dear Ladies and Gentlemen. If you are writing to personnel within a police department, you will want to address the person: e.g., Dear Sergeant Chung, Dear Chief Stevens, or Dear Constable Hassan.

Subject Lines

Subject lines are optional. You do not need to use them, but you may find that a subject line comes in handy if you have written a long letter. It will give the reader the chance to quickly glance at the subject line to get an idea of what the letter is about before she reads the letter. If you do use one, underline the entire subject line.

Body

This is the actual message you want to send to your reader. It consists of a minimum of three paragraphs.

Complimentary Close

The complimentary close puts an end to your letter. Common phrases such as "Yours truly" or "Sincerely yours" are used as complimentary closes.

Identification Line and Signature

The identification line consists of your typed name and title. Place your signature between the complimentary close and identification line.

Enclosure

The abbreviation "Enc." is used at the bottom of the letter if you are enclosing something with the letter.

BODY STRUCTURE OF THE LETTER

The structure of the body of the letter is similar to the memo. You want to make your point quickly. The reader wants to know your main message quickly. You can then proceed with your details and explanations. Follow the indicated structure for each paragraph. As with the memo, the amount of information you provide in your body paragraphs will depend on your subject and how much information you need to give your reader.

Introductory Paragraph

You want to give your reader the most important message here. Tell the reader the most important point you need to convey. Determine the main message of the letter and then place the main message in the first paragraph. You may need to lead into your main message, but it should appear in the first paragraph.

Middle Paragraph(s)

In the middle paragraph(s) you will need to give your reader the details and explanations. Think about everything that the reader will want to know and needs to know. The information you provide in your body paragraphs needs to back up your main message. To complete this information, ask yourself the five Ws — *who*, *what*, *when*, *where*, and *why*, as well as *how*. Answering these questions will help you provide complete information.

Concluding Paragraph

The concluding paragraph of a letter differs from the concluding paragraph of a memo. There are several things you want to include in this paragraph. The first thing you should do is repeat your most important message. Also, before closing, look to the future, end in a positive way, and give the reader a contact number. Often when you are using company letterhead, a phone number is already indicated. All you may have to add is an extension number where you can be reached.

SKETCH OF A LETTER

Anywhere Police Department
100 Anywhere Street
City, Province
X4R 3E2

March 2, 2006

Metropolitan City Police Department
123 Anywhere Street
City, Province
P9M 7Y5

Dear Constable Rajinder Patel:

Subject:_____

GIVE YOUR MAIN MESSAGE

DETAILS AND EXPLANATIONS

DETAILS AND EXPLANATIONS

REPEAT MAIN MESSAGE
END IN A POSITIVE WAY
LOOK TO THE FUTURE
GIVE A CONTACT NUMBER OR EXTENSION

Yours truly,

Constable Susan Lynch
SL/at

PRACTICE: LETTER PARAGRAPH STRUCTURE

What information should be provided in each paragraph of the letter in each of the following situations?

1. Ordering new uniforms for new police officers

_____

_____

2. Inviting community members to sit on a committee to improve neighbourhood watch

_____

_____

3. Accepting an invitation from a local elementary school to do a road-safety presentation for schoolchildren

_____

_____

4. Inviting a special guest to be the speaker for the annual Police Appreciation Banquet

_____

_____

5. Requesting more information about a computer-training workshop for police officers

_____

_____

LETTER STYLES

There are four main styles you can choose from. They are the full-block, semi-block, simplified, and modified-block styles.

Full-Block Style

This style is probably the most popular because it is the easiest to use. What you need to remember about this style is that everything is typed at the left margin. You do not need to worry about centring anything. Your return address, date, inside address, salutation, subject line, body paragraphs, complimentary close, identification line, and enclosure, if you are using one, all start at the left margin. Remember that if you are using this style, your body paragraphs are not indented.

Semi-Block Style

In this style, you centre the return address (if you are writing a personal letter), date, complimentary close, and identification line. If you are using a subject line and an enclosure, they should also be centred. Indent the first line of your body paragraphs five spaces.

Simplified Style

This style differs from the others in that it has no salutation or complimentary close. As with the full-block style, all the information is typed flush against the left margin. Your return address (if you are not using letterhead), date, inside address, subject line, body paragraphs, identification line, and enclosure are typed at the left margin. Do not indent the first line of your body paragraphs.

Modified-Block Style

In this style, the return address, date, complimentary close, and identification line are typed on the side of the page to the right of the centre. The inside address and salutation are typed at the left margin. Body paragraphs are not indented; they are typed flush against the left margin.

Punctuation

There are two different punctuation styles that you can use: two-point or open punctuation. **Two-point punctuation** involves placing a colon after your salutation and a comma after your complimentary close. **Open punctuation** avoids the use of punctuation after the salutation and complimentary close. As the name implies, these are left open.

Open punctuation is the preferred style because it is easier. You don't have to worry about whether or not you have placed the correct punctuation in the proper place. These two punctuation styles are also flexible. That is, they don't belong with one particular letter style. You can use open or two-point punctuation with the full-block style. It doesn't matter which punctuation style you choose, as long as you apply it consistently.

EXAMPLE OF FULL-BLOCK STYLE

Somewhere Police Department
567 Downtown Avenue
Somewhere, Ontario
J7T 5R3
(221) 513-9245

March 2, 2006

Metropolitan City Police Department
123 Anywhere Street
City, Province
P9M 7Y5

Dear Sergeant Kranston:

Re: Conducting Computer Workshop

The Somewhere Police Department is becoming completely computerized, and all reports will be computer-generated. The officers have requested some training in this area to prepare for the changes ahead. We are aware that you are an expert in this field and would greatly appreciate your assistance in facilitating a series of workshops for our officers.

If you decide to accept our offer, all your expenses will be paid in full. We will pay for your accommodations, travel, and meals. We will also pay for your efforts in conducting the workshops. If possible, we would appreciate your facilitating workshops during the week of April 17-21. We would set up a series of workshops to enable all officers to attend one of the sessions.

Again, we would be interested in having you lead a series of workshops for our officers related to the new computer system. I look forward to hearing from you. You can reach me at extension 777.

Yours truly,

Sergeant Matthew O'Brien
MO/tt

Note the two-point punctuation in this letter.

Somewhere Police Department
567 Downtown Avenue
Somewhere, Ontario
J7T 5R3
(221) 513-9245

March 2, 2006

Metropolitan City Police Department
123 Anywhere Street
City, Province
P9M 7Y5

Dear Sergeant Kranston

RE: Conducting a Computer Workshop

The Somewhere Police Department is becoming completely computer-ized, and all reports will be computer-generated. The officers have requested some training in this area to prepare for the changes ahead. We are aware that you are an expert in this field and would greatly appreciate your assistance in facilitating a series of workshops for our officers.

If you decide to accept our offer, all your expenses will be paid in full. We will pay for your accommodations, travel, and meals. We will also pay for your efforts in conducting the workshops. If possible, we would appreciate your facilitating workshops during the week of April 17–21. We would set up a series of workshops to enable all officers to attend one of the sessions.

Again, we would be interested in having you lead a series of workshops for our officers related to the new computer system. I look forward to hearing from you. You can reach me at extension 777.

Yours truly

Sergeant Matthew O'Brien
MO/tt

Note the open punctuation in this letter.

EXAMPLE OF SIMPLIFIED STYLE

Somewhere Police Department
567 Downtown Avenue
Somewhere, Ontario
J7T 5R3
(221) 513-9245

March 2, 2006

Metropolitan City Police Department
123 Anywhere Street
City, Province
P9M 7Y5

RE: Conducting Computer Workshop

The Somewhere Police Department is becoming completely computerized, and all reports will be computer-generated. The officers have requested some training in this area to prepare for the changes ahead. We are aware that you are an expert in this field and would greatly appreciate your assistance in facilitating a series of workshops for our officers.

If you decide to accept our offer, all your expenses will be paid in full. We will pay for your accommodations, travel, and meals. We will also pay for your efforts in conducting the workshops. If possible, we would appreciate your facilitating workshops during the week of April 17–21. We would set up a series of workshops to enable all officers to attend one of the sessions.

Again, we would be interested in having you lead a series of workshops for our officers related to the new computer system. I look forward to hearing from you. You can reach me at extension 777.

Sergeant Matthew O'Brien

MO/tt

Somewhere Police Department
567 Downtown Avenue
Somewhere, Ontario
J7T 5R3
(221) 513-9245

March 2, 2006

Metropolitan City Police Department
123 Anywhere Street
City, Province
P9M 7Y5

Dear Sergeant Kranston:

Subject: Conducting Computer Workshop

The Somewhere Police Department is becoming completely computerized, and all reports will be computer-generated. The officers have requested some training in this area to prepare for the changes ahead. We are aware that you are an expert in this field, and would greatly appreciate your assistance in facilitating a series of workshops for our officers.

If you decide to accept our offer, all your expenses will be paid in full. We will pay for your accommodations, travel, and meals. We will also pay for your efforts in conducting the workshops. If possible, we would appreciate your facilitating the workshops during the week of April 17–21. We would set up a series of workshops to enable all officers to attend one of the sessions.

Again, we would be interested in having you lead a series of workshops for our officers related to the new computer system. I look forward to hearing from you. You can reach me at extension 777.

Yours truly,

Sergeant Matthew O'Brien
MO/tt

Note the two-point punctuation in this letter.

TONE

As with memos, when you are writing letters, remember to use a positive tone. Your message will be received more favourably and it will keep the reader with you.

FOCUS ON *YOU*

Again, as with memos, when you are writing letters, use a *you* focus. Keeping with this focus will help you to emphasize the benefits of the message to the reader.

PRACTICE: TONE AND *YOU* FOCUS

Read the following letter and make changes so that the tone is positive and that a *you* focus is apparent.

Metropolitan Police Department
123 Somewhere Avenue
Somewhere, Ontario
P9J 4S7
(765) 865-3420

November 10, 2006

Sergeant Milton Bringleson
Anywhere Police Department
129 Anywhere Street
Anywhere, Ontario
Y6T 3F4

Dear Sergeant Milton:

I need you to help me. I am looking for information on employee absences. I know that you have policies and procedures in place that deal with this issue, and I need your information so that I can put one together.

There have been many problems at our station with some of our officers taking too many days off. I don't know what the problem is with some of these people, but they think it is okay to take a day off when they feel like it. You have dealt with this problem and can tell me what to do.

Send me your policies and procedures as soon as possible. If you need me, I am at extension 876.

Sincerely yours,

Sergeant Kayleen Tossles

LETTER FORMAT: WITH LETTERHEAD

If you are using letterhead, set up your letter in the following way:

Metropolitan Police Department
123 Anywhere Street
Anywhere, Ontario
P9J 4S7

Date   → Begin your date on line 13.
   ↓ Leave 2 to 9 blank lines depending on the length of your letter.

Inside Address
   ←
   Leave 2 blank lines.
Salutation
   ←
   Leave 1 blank line.
Subject Line
   ←
   Leave 1 blank line.

Body (Leave 1 blank line between
        paragraphs. Information contained
        within a paragraph is single spaced.)

   1 blank line
Complimentary Close

        Leave 4 to 5
        blank lines.

Your typed name
Title
   ← 1 blank line
LETTER WRITER'S INITIALS/typist's initials
   ← 1 blank line
Enc.

LETTER FORMAT: PERSONAL LETTERS

If you are sending a personal letter, set up your letter in the following way:

Sender's Address   →   Begin on line 7.

←   Leave 1 blank line.

Date

←   Leave 2 to 9 blank lines depending on
      the length of your letter.

Inside Address

←   Leave 2 blank lines.

Salutation

←   Leave 1 blank line.

Subject Line

←   Leave 1 blank line.

Body (Leave 1 blank line between
        paragraphs. Information contained
        within a paragraph is single spaced.)

1 blank line
Complimentary Close

        Leave 4 to 5
        blank lines.

Your typed name
Title

←   1 blank line

Letter Writer's Initials/typist's initials

←   1 blank line

Enc.

STAGES INVOLVED IN WRITING THE LETTER

Letter writing involves the prewriting, writing, and rewriting stages.

Prewriting Stage

1. Make a list of the points you will include in your letter.
2. Review your list and cross off any points that you will not need. Place a check mark next to the points you will keep.
3. Fill in your outline.

Writing Stage (Rough Draft)

Keep your outline at your side and begin writing. Don't worry about sentence structure or grammatical errors at this point. The idea is to get all your ideas on paper in sentence form.

Rewriting

Once the rough draft is complete, you can begin looking at your letter for revisions. As you are editing, ask yourself these questions:

1. Are all the parts of the letter included?
2. Is the style of the letter consistent and correct?
3. Is the two-point or open punctuation used correctly?
4. Is the most important message in the introductory paragraph?
5. Does the body paragraph contain the details and explanations?
6. Does the concluding paragraph repeat the main message, look to the future, end in a positive way, and give a contact number?
7. Is the letter accurate and complete?
8. Are all of the reader's questions answered?
9. Are there any grammatical problems that need to be cleaned up?

Keep editing and rewriting until you are satisfied that you have a copy you are comfortable sending.

OUTLINE FOR WRITING THE LETTER

Date: _____

Addressee's Name _____
Addressee's Title_____
Full Address _____
City/Province _____
Postal Code_____

Dear _____

Subject:_____

Introductory Paragraph:
Main Message

_____

_____

_____

Body Paragraph:
Details and Explanations

_____

_____

_____

_____

Concluding Paragraph:
Repeat the main message _____
Look to the future_____
End in a positive way _____
Offer a contact number or extension _____

Complimentary Close_____

Your name_____
Title_____

YOUR INITIALS/typist's initials

SAMPLE WRITING PROCESS FOR THE LETTER

Here is the process one constable went through to complete a letter.

Lisa Phillips is a constable working at the Metropolitan Police Department. She has just received a letter from Joel Fisher, dean of criminal justice programs at Community College. The letter asks her to join the advisory committee for the Law and Security Program. She is delighted and honoured to be asked and is definitely interested in accepting the invitation. Lisa decides to accept the invitation by responding with a letter. Here are the steps Constable Phillips went through to complete the letter:

1. Making a List

- surprised by the invitation
- great honour to have been asked to become a member of the advisory committee
- graduate from the law and security program and can contribute to the discussions
- will benefit the board because of extensive knowledge in the policing area
- 15 years of policing will come in handy on the committee
- have had experience teaching a course in the law and security program — criminal justice procedures
- will be able to inform the advisory board about what the policing profession is currently doing and what is expected from new recruits
- happy to do this extra work even though there is no money involved
- chief will be excited to hear that I was asked
- definitely accept invitation
- contact me with dates and times of the meetings

2. Checking off Ideas

Once Constable Phillips listed the ideas she thought she would include in her letter, she reviewed and checked off the ones she would keep and placed an *X* beside the ideas she felt were not relevant. Here is how her list looked:

- surprised by the invitation ✗
- great honour to have been asked to become a member of the advisory committee ✓
- graduate from the law and security program and can contribute to the discussions ✗
- will benefit the board because of extensive knowledge in the policing area ✓
- 15 years of policing will come in handy on the committee ✓
- have had experience teaching a course in the law and security program—criminal justice procedures ✓
- will be able to inform the advisory board about what the policing profession is currently doing and what is expected from new recruits ✓
- happy to do this extra work even though there is no money involved ✗
- chief will be excited to hear that I was asked ✗
- definitely accept invitation ✓
- contact me with dates and times of meetings ✓

3. Filling in the Outline

Once Constable Phillips had checked off the ideas she thought would be appropriate, she made an outline to help her begin writing. Here is what her outline looked like. Notice that she did not write full sentences but used point form.

OUTLINE FOR WRITING THE LETTER

Date: September 22, 2006

Addressee's Name   Joel Fisher
Addressee's Title   Dean, Criminal Justice Programs, Community College
Full Address   1234 Anywhere Street
City/Province   Anywhere, Ontario
Postal Code   P1Q 2R3

Dear Mr. Fisher:

Subject: _____

Introductory Paragraph:

Main Message
— accept invitation to be a member of the advisory board for Criminal Justice Programs

Body Paragraph:

Details and Explanations
— understand Law & Security Program has excellent reputation—extensive policing experience
— 15 years in force—taught Criminal Justice Procedures in 2004
— will be an asset to board

Concluding Paragraph:

Repeat the main message — gladly accept invitation

Look to the future — look forward to working with you

End in a positive way

Offer a contact number or extension — extension 435

Complimentary Close   Sincerely yours,

Your name
Title

YOUR INITIALS/typist's initials

4. Writing the Rough Draft

After Constable Phillips completed her outline, she was ready to begin writing her rough draft. Here is how her first draft looked:

Metropolitan Police Department
5678 Somewhere Avenue
Anywhere, Ontario
P9J 4S7
(888) 235-9876

September 22, 2006

Joel Fisher
Dean, Criminal Justice Programs
Community College
1234 Anywhere Street
Anywhere, Ontario
P1Q 2R3

Dear Mr. Fisher:

Thank you for the invitation to sit on the Advisory Board for the Law and Security Program at Community College. I understand that the Law and Security Program has an excellent reputation.

I would be a benefit to this committee for many reasons. I have extensive policing experience, having worked for 15 years as a constable here at Metropolitan Police Department. I have also had the opportunity to teach the course Criminal Justice Procedures at Community College in 2004. This was an opportunity I enjoyed tremendously. I must say the students were a joy to teach. Because of my extensive policing experience and my teaching experience, I feel I will be a benefit to the Board. I gladly accept your invitation. If possible, please contact me and let me know when the Board meets so I can make arrangements with my schedule.

Once again, I'm honoured to be on the Advisory Board for the Law and Security Program at Community College. I look forward to working with you. You can reach me at extension 435.

Sincerely Yours,

Constable Lisa Phillips

---

5. Editing the Rough Draft

Once Constable Phillips had her rough draft completed, she edited her work and made the changes she thought necessary. Look at the changes she made.

Metropolitan Police Department
5678 Somewhere Avenue
Anywhere, Ontario
P9J 4S7
(888) 235-9876

September 22, 2006

^Mr. Joel Fisher
Dean, Criminal Justice Programs
Community College
1234 Anywhere Street
Anywhere, Ontario
P1Q 2R3

Dear Mr. Fisher:

become a member of the

Thank you for the invitation to ~~sit on the~~ Advisory Board for the Law and Security Program at Community College. ~~I understand that the Law and Security Program has an excellent reputation.~~ *not relevant*

I would be a benefit to this committee for many reasons. I have extensive policing experience, having worked for 15 years as a constable here at Metropolitan Police Department. I have also had the opportunity to teach the course Criminal Justice Procedures at Community College in 2004. This was an opportunity I enjoyed tremendously. ~~I must say the students were a joy to teach.~~ Because of my extensive policing experience and my teaching experience, I feel I will be a benefit to the Board. I gladly accept your invitation. If possible, please contact me and let me know when the Board meets so I can make arrangements <u>with my schedule.</u> *wording- arrangements to schedule shifts around meetings*

move to first paragraph – main idea

Once again, I'm honoured to be on the Advisory Board for the Law and Security Program at Community College. I look forward to working with you. You can reach me at extension 435. *with further details*

Sincerely Yours,

Constable Lisa Phillips

6. Completing the Final Copy

Constable Phillips wrote, edited, and rewrote until she was satisfied that she had a copy she could mail. Take a look at her final version.

Metropolitan Police Department
5678 Somewhere Street
Anywhere, Ontario
P9J 4S7
(888) 235-9876

September 22, 2006

Mr. Joel Fisher
Dean, Criminal Justice Programs
Community College
1234 Anywhere Street
Anywhere, Ontario
P1Q 2R3

Dear Mr. Fisher:

Thank you for the invitation to become a member of the Advisory Board for the Law and Security Program at Community College. I gladly accept your invitation.

Because of my policing background and my teaching experience, I feel I will be a benefit to the Board. I have extensive policing experience, having worked for 15 years as a constable here at Metropolitan Police Department. I am familiar with policing practices and what the policing profession is expecting from the graduate and potential police officer. I have also had the opportunity of teaching the Criminal Justice Procedures course at Community College in 2004, so I am familiar with the policing curriculum. If possible, please contact me and let me know when the Board meets. I would like to make arrangements so that my shifts are scheduled around the Board meeting times.

Once again, I accept the invitation to sit on the Advisory Board for the Law and Security Program at Community College. I look forward to working with you. You may reach me at extension 435 with further details.

Sincerely yours,

Constable Lisa Phillips

PRACTICE: COMPLETING THE LETTER

You are given the introductory paragraph of the letter. Complete the letter by writing the middle paragraph and the closing paragraph.

Metropolitan Police Department
5678 Somewhere Street
Somewhere, Ontario
P9J 4S7

May 18, 2006

Sergeant Sam Jones
Anywhere Police Department
129 Anywhere Street
Anywhere, Ontario
P2M 8T5

Dear Sergeant Jones:

Thank you very much for your invitation to become a member of the Provincial Policing Safety Committee. I gladly accept your invitation.

_____

_____

_____

_____

_____

_____

_____

Yours truly,

Sergeant Lydia Macintosh

PRACTICE: COMPOSING THE LETTER

Sergeant Murphy needs to send a letter to Sergeant Liddens of the Somewhere Police Department to obtain information about his department's report-writing procedures. Sergeant Murphy is interested in knowing how the officers at the Somewhere Police Department complete their police reports. Sergeant Murphy has reached the point where he has listed possible ideas that would be included in his letter. Review his list and check off the ideas you think should go in the letter. Cross off the ideas you believe are not necessary. You may also want to add ideas that you feel are missing. Then go through the remaining stages of the writing process and finish Sergeant Murphy's letter.

- officers here at the Metropolitan Police Department are inconsistent when it comes to writing police reports
- some officers use the third person, others use the first person
- some officers use complete sentences when they are dictating reports; others do not
- some officers have a problem with including subjective comments in their police reports
- these inconsistencies are causing problems within the department
- the support staff are complaining because the typists say this makes their jobs difficult
- sometimes I think they are just trying to find something to complain about
- I would like to know from you what your procedures are
- in particular, I am interested in knowing whether you use third or first person, whether your officers dictate in complete sentences, and how you handle the issue of subjective comments in police reports
- let me know right away what you think
- I need to do something about this situation soon

DIFFERENT TYPES OF POLICE LETTERS

As a police officer, you will encounter a variety of situations that require you to write different types of letters. The most common letter styles that you will use are the suggestion letter, request letter, information letter, confirmation letter, and responding to a complaint letter. These five letters follow the basic structure for letter writing. Overall, you need to remember that the main message goes in the first paragraph, details and explanations belong in the middle paragraphs, and that in the concluding paragraph you need to repeat the main message, look to the future, end positively, and offer an extension number. On the following pages, you will find a more detailed version of each of these five types of letters and what needs to go in each paragraph. You need to remember that in these five different types the details and explanations go together. That does not necessarily mean that you need only one body paragraph. You must judge whether the ideas belong together or in separate paragraphs.

On the following pages you will find explanations of the structure of each type of police letter, followed by an example of each. Look them over to get an idea of their basic structure.

Suggestion Letter

- Make the suggestion.

- Include the reason for the suggestion as well as any background information.
- Explain why the idea will be useful.
- Explain what the idea will achieve.

- Briefly repeat the suggestion.
- Tell the reader what she needs to do.
- Look to the future/end positively.
- Provide a phone number/extension.

Request Letter

- Make your request.

- Explain why you want it.
- Explain what the benefits are.
- Give a detailed description of the request (any costs, what it involves).

- Repeat your request.
- Tell your reader what she needs to do.
- End positively/look to the future.
- Provide a phone number/extension.

Information Letter

- State the main message.

- Explain the circumstances or reference to a request.
- Give the facts to support or explain the main message.

- Briefly repeat the main message.
- Tell the reader what action she needs to take.
- Look to the future/end positively.
- Provide a phone number/extension.

Confirmation Letter

- Briefly state that you are confirming something.

- Refer to the circumstances and any previous correspondence.
- Clearly define the facts and figures that both parties have agreed to.

- Briefly restate the main message.
- End positively/look to the future.
- Tell the reader what she needs to do.
- Provide a phone number/extension.

Metropolitan Police Department
5678 Somewhere Street
Anywhere, Ontario
P9Y 4S7
(888) 235-9876

October 23, 2006

Mack Simpson
Manager, Special Events
129 Anywhere Street
Anywhere, Ontario
P1Q 2R3

Dear Mr. Simpson:

Thank you for your letter informing us of the Spring Fever Pub that will be held at your arena on February 15, 2007. It is helpful for us to know that an event of this magnitude will be held in our city. However, I would like to make one suggestion. Because the event is of a large capacity, I suggest that four uniformed police officers be present to ensure that no problems arise.

The arena holds about 1500 people. This is a large crowd to have drinking without some kind of safety precaution. Uniformed officers will reduce the chance of the crowd getting out of control. Having officers present at the pub will also be helpful if an emergency situation does occur. The total cost for having this service present at the pub is $1652.00. If you choose this option, I am sure you will find that this is money worth spending.

Again, I would recommend that four uniformed officers be present at the Spring Fever Pub. You may reach me at extension 546. Please feel free to call me if you have any questions or concerns. I look forward to hearing from you.

Sincerely yours,

Sergeant Dwayne Hodges

DH/st

REQUEST LETTER

Metropolitan Police Department
5678 Somewhere Street
Anywhere, Ontario
P9Y 4S7
(888) 235-9876

March 6, 2006

Ms. Derinda Delouse
Principal
Anywhere Public School
456 Anywhere Street
Anywhere, Ontario
P1Q 2R3

Dear Ms. Delouse:

Over the last few months, various constables in the department have been touring the elementary schools in the area conducting safety presentations for the children. We would at this time like to request the opportunity to visit your grade one, two, three, and four students.

Topics we will cover include road safety and bike safety. Of course, we will bring along Sebastian the safety alligator for students' enjoyment. Our safety program has been a successful one, and we hope to share this experience with your students. Our presentation takes about 50 minutes. This includes time for the children to ask questions and spend some time with Sebastian. Parents are also more than welcome to attend.

I would very much appreciate the opportunity to speak with the students. I will contact you in the near future to discuss this opportunity. I look forward to talking with you. If you have any questions or concerns, feel free to contact me at extension 345.

Sincerely yours,

Sergeant Michelle Evans

Metropolitan Police Department
5678 Somewhere Street
Anywhere, Ontario
P9J 4S7
(888) 235-9876

March 6, 2006

Mayor Melinda Moffat
423 Anywhere Street
Anywhere, Ontario
P9Y 3R4

Dear Mayor Moffat:

The Metropolitan Police Department has just completed a two-month Road Safety Check. The program was successful, showing improvements from last year's campaign.

Over the last two months, constables randomly stopped 300 drivers and checked for valid drivers' licences and insurance. A seat belt check was also completed on each car. In total, only 35 drivers were charged with some violation. These consisted of 15 with invalid licences, 12 with no insurance, and 8 with seat belt violations. The overall numbers are down from last year's campaign, in which 20 were charged with driving without a valid licence, 14 without insurance, and 10 with seat belt violations.

Our campaign this year was a success. These numbers are a reassurance that our program to promote road safety is working. If you would like further information or details, please feel free to contact me at extension 789. I look forward to hearing from you.

Sincerely,

Chief Barney Bijowski

Metropolitan Police Department
5678 Somewhere Street
Anywhere, Ontario
P9Q 4S7
(888) 235-9876

March 6, 2006

Sergeant Michael Holmes
Valadero Police Services
1224 Valadero Street
Valadero, Ontario
P8Y 6T4

Dear Sergeant Holmes:

I am looking forward to attending this year's police conference in Valadero, Ontario. The conference sounds as if it has a lot to offer police officers. I am happy to be confirming that I, along with two other officers, will be offering the self-defence workshop at the conference.

Our workshop will provide the participants with an overview of some self-defence techniques that they may find useful in certain conflict situations. The workshop will provide an overview of what the various techniques are, when they are best used, and will also provide a demonstration of each technique. Participants will be given the opportunity to practise the techniques demonstrated.

Again, we are happy to be providing a workshop for the annual police conference. We look forward to participating and working with other officers from around the province. If you have any questions or concerns about our presentation, feel free to contact me at extension 659.

Sincerely yours,

Sergeant Glenda McLaughlin

Responding to a Complaint Letter

- Briefly agree or disagree that the claim or complaint is valid.

- Refer to the reader's original letter of request and explain why you agree or disagree.
- Restate the most important idea.

- State what you would like the reader to do.
- Look to the future/end positively.
- Provide a phone number/extension.

RESPONDING TO A COMPLAINT LETTER

Metropolitan Police Department
5678 Somewhere Street
Anywhere, Ontario P1Q 2R3
(888) 235-9876

March 6, 2006

Ms. Christina Mackeral
97 Mystery Street
Anywhere, Ontario
P9T 6R4

Dear Ms. Mackeral:

Re: Investigation # 941054

I received your complaint regarding the handling of the car accident report by Constable Miners and have investigated it further. I have come to the conclusion that the case was handled properly.

Under Section 000 of the *Canadian Criminal Code*, losing control of a motor vehicle is not classified as careless driving. Careless driving is intentionally violating safety precautions while in control of a motor vehicle. Mr. Morrison was not driving at a high speed, swaying from side to side, ignoring traffic signs or lights, or exhibiting any other dangerous behaviour. There was black ice where he lost control of his car, and that is what caused the accident.

Although unfortunate, your accident was entirely a mishap due to weather and road conditions. I assure you all proper procedures were taken in the investigation of your accident. You are welcome to come to my office to discuss the details of the investigation. I can be contacted at extension 246.

Sincerely,

Chief Vonnie Vonderburg

USING COMPUTER TEMPLATES TO WRITE LETTERS

As with memos, preformatted letter templates also exist within some software. Some letter templates are preformatted so that you can fill in the return address, date, salutation, and complimentary close. You do not have to worry about forgetting parts of the letter. The blanks are there for you to simply fill in. Before you use a preformatted letter template, ensure that it meets your needs. Most organizations have company letterhead and all correspondence must be completed on the letterhead.

Using Grammar and Spelling Checkers

With memos, we discussed using grammar and spelling checkers. The same rules would apply to letter writing. The grammar and spelling checkers are good to use as a resource, but as mentioned earlier, do not totally rely on them. Run through the grammar and spelling checker at least twice, but also review your document from a hard copy and check for grammar and spelling errors on your own. Also, keep in mind that sometimes the grammar and spelling checkers will recognize something in your text as an error that may not be an error, or it may suggest making a correction that really does not correct the problem. Be aware that this may happen and give yourself enough time to go over the errors identified to decide if they really are errors and determine how to make the correction. If you are unsure of an error and how to correct it, consult reference books, dictionaries, or ask someone to help you edit your document.

PRACTICE: WRITING POLICE LETTERS

Write a letter for each one of the scenarios below.

Suggestion Letter

A. You are a staff sergeant working at the Metropolitan Police Department. You know that the Busytown Police Services is holding this year's annual police conference. You have attended these conferences for the past five years and would like to see more workshops made available for officers that deal with self-defence techniques. Write to Staff Sergeant Hari Prashad at the Busytown Police Services to make the suggestion that more self-defence workshops be held at this year's conference.

B. You are Chief Boudreau working at the Anytown Police Department. You have recently returned from the biannual chiefs' meeting. You feel the meetings could take place through teleconferencing instead of travelling to a different location each time a meeting is to occur. You believe it would save each police department a considerable amount of money. Write a letter to Chief Chen, chair of the Chiefs Committee, making this suggestion.

Request Letter

A. You are a constable working at the Metropolitan Police Department. Chief Sobel has asked you to compile information on drug use by youth in the Metropolitan area. The police force recognizes that drug use is a growing problem, but has no documented research on how bad the situation really is. With the information you gather, you hope to develop drug prevention and education strategies. Chief Sobel tells you that the Youth Alternative Centre is a good place to start to locate information. The Youth Alternative Centre is a place where young people seek help for drug rehabilitation. You decide to write to Samantha Lethons, director of the Youth Alternative Centre, to request any available information regarding this problem. You are particularly interested in information regarding the age groups using drugs and what types of drugs they are using. You also want to ask Ms. Lethons for suggestions on how to prevent this problem.

B. You are a staff sergeant working at the Metropolitan Police Department. Chief Neveau has asked you to investigate the cost of installing computers in the cruisers to help officers with their report writing. Write ABC Computers a letter requesting prices for this endeavour.

Information Letter

A. You have received some excellent information from Samantha Lethons, director of the Youth Alternative Centre, regarding the types of drugs youths seem to be using in the area and what age groups seem to be using the most. Ms. Lethons also provided you with some recommendations on how to resolve this growing problem. Since then, the police department has taken steps to rectify the drug problem in the metropolitan area. Some of these steps include having officers visit local high schools to give students a presentation on the drug problem, patrol high-risk areas more frequently, and become involved in the drug-awareness week program in the community. Write to Ms. Lethons outlining some of the initiatives the police force has taken.

B. You are a sergeant working at the Anytown Police Department. You have just returned from attending a two-day computer workshop at another police station. You enjoyed the workshop and learned a lot. Write Sergeant Bleu a letter informing him how much you enjoyed and benefited from this workshop.

Confirmation

A. You are a sergeant working at the Metropolitan Police Department. You have been approached by the Addiction Research Centre and asked if the police department would become involved in the annual Drug Awareness Week. You believe this is a good cause and that your officers would like to participate. Write a letter to Larry Suleman, director of the Addiction Research Centre, confirming your staff's participation in this year's Drug Awareness Week. In the letter, explain the event that the police department plans to hold to promote the "Say No to Drugs!" campaign.

B. You are Chief Garnett working at the Metropolitan Police Department. You have been invited by the Anytown Police Department Board to attend their next meeting. They would like to get some information from you concerning your department's policies. You would be happy to attend their meeting and provide them with some information. Write Chief Levesque a letter confirming your attendance at their next meeting.

Response to a Complaint

A. You are a sergeant working at the Metropolitan Police Department. You have received a complaint from Julia Mayworth about how one of your constables handled a noise complaint. Ms. Mayworth had called the police complaining about a neighbour playing loud music. Constable Jackson was dispatched to this complaint but did not lay any charges. Ms. Mayworth feels the constable mishandled the situation. You, as the sergeant, looked into the matter and feel the situation was handled properly. Write Julia Mayworth a letter responding to the complaint.

B. You are Chief Lethbridge working at the Metropolitan Police Department. You received a complaint that one of your officers, Constable Mellows, used excessive force when dealing with a domestic call. You investigated the situation and feel that Constable Mellows was justified in her actions. Write a response to the complaint you received, explaining the outcome of your investigation into this matter.

DISCUSSION QUESTIONS

1. What is the purpose of the letter?
2. Describe the four letter styles and how they differ.
3. What is the difference between open punctuation and two-point punctuation?
4. What information should be included in each paragraph of the letter?
5. Describe the different types of police letters.

ACTIVITY: EVALUATING LETTERS

Evaluate the following letters. Correct any errors in format, paragraph structure, grammar, and wording. Every letter should be in full-block format, and the punctuation should be consistent within each letter.

Metropolitan Police Department
5678 Somewhere Street
Anywhere, Ontario
P9Q 4S7
(888) 235-9876

March 6, 2006

Janice Brown
4321 Oak Grove Drive
Mystery, Ontario
Q3W 6T7

Dear Ms. Brown:

I am writing this letter confirming our previous tentative agreement. In case you do not recall our agreement it is to have myself on a committee looking at the use of pepper spray.

I received your previous letter on January 16, 2006, and an eager to participate in this new committee to look at the regulations surrounding pepper spray. I am able to attend to meeting on March 27, 2006, and will be willing to put in as much input as possible.

Again, I am just confirming our previous agreement to be part of a provincial committee looking at the regulations surrounding the use of pepper spray. When you receive this letter, please call me at (906) 234-5679.

Sincerely Yours

Constable Amanda Davies

Metropolitan Police Department
5678 Somewhere Street
Anywhere, Ontario
P9Q 4S7
(888) 235-9876

March 6, 2006

Carl Mostoff
1209 Mystery Street
Anywhere, Ontario
P9T 6R4

Dear Mr. Mostoff:

With regards to your complaint, I would have to disagree. The accident was investigated properly, and no one was at fault for the accident.

Therefore, I disagree with your complaint. If you have any questions, let me know.

Thank you

Chief Luigi Mastroni

Metropolitan Police Department
5678 Somewhere Street
Anywhere, Ontario
P9Q 4S7
(888) 235-9876

March 6, 2006

Mr. John Doe
252 Mystery Street
Anywhere, Ontario
P8Y 5E3

My name is Sergeant Leonard Farr of the Metropolitan Police Department. As you may be aware of from the local newspapers or victim of the increase in break ins in the local area of Michigan Avenue and Walnut Street. As being police officers, one of our jobs is to prevent break and enter crimes from happening to your home. We also investigate and pursue any suspects that commit break and enters.

The Metropolitan Police Department will be increasing patrol to the above stated area to help prevent anymore break and enters from happening and hopefully catch the person or persons involved in this crime. I still strongly suggest that everyone should keep there belongings locked up in either the house or garage. Also keep you doors and windows locked. Even if you are going down to the corner street.

If you have any questions, feel free to call me at the extension 754.

Yours Truly,

Leonard Farr

Sergeant

Metropolitan Police Department
5678 Somewhere Street
Anywhere, Ontario
P9Q 4S7
(888) 235-9876

March 6, 2006

Mr. Harry Harrison

23 Mystery Street Anywhere, Ontario

P9Y T5W

Mr. Harrison:
I am writing this letter in response to the formal complaint you filed.

In your complaint you wrote that the roads were to badly covered with snow for any person to see the merging lanes. At the seen I remember you making that same statement. I could see the lines quite fine when I checked your license it clearly stated that you must be wearing your glasses at all times or your license in invalid.

You were not wearing your glasses at the seen of the accident, and you could not produce them when asked. Your complaint has been heard and turned down. If you have any more comments, please contact me at 567-9876, extension 124.

Thank you

Brenda Salguero

Constable

5678 Somewhere Street
Anywhere, Ontario
P9Q 4S7
(888) 235-9876

March 6, 2006

Dear Ms. Lymboski

I am writing to you to request that you allow myself and various staff members of my police service to visit and make a safety presentation to the children of your school. We would like to promote safety to the children because some children are unaware of the dangers in society. This will provide the children with the knowledge of safety on the streets, at school, and at home. We would like to talk to the children between the grades of one to four. This includes the children as well as the staff participating in games, as well as many discussion groups. The cost is minimal, therefore, the police department will cover it. It would be beneficial for the children of your school, if you would allow myself and fellow officers to make a safety presentation. Please contact me at the station as soon as possible. I am looking forward to hearing from you soon. I can be reached at (888) 235-9876, extension 2698.

Sincerely yours,

Sergeant Tim Jones

Metropolitan Police Department
5678 Somewhere Street
Anywhere, Ontario
P9Q 4S7
(888) 235-9876

March 7, 2006

Mrs. Melanie Mackinnon
Brule Public School
234 Somewhere Street
Anywhere, Ontario
P1Q 2R3

Dear Mrs. Mackinnon

RE: School Safety Programs
Our department has promoted safety many times throughout the community.
We would like to expand this program into the local elementary schools.

We would really like to further this program into the school, if you are interested please contact me at (888) 235-9876, ext. 486. I look forward to hearing from you.

Sincerely,

Sergeant Steven Scott

ACTIVITY: EVALUATING TONE

Evaluate the tone of the following letter. How could it be rewritten to improve the tone?

Metropolitan Police Department
5678 Somewhere Street
Anywhere, Ontario
P9Q 4S7
(888) 235-9876

March 31, 2006

Mr. John Garrison
1914 Somewhere Street
Anywhere, Ontario
P1Q 2R3

Dear Mr. Garrison:

I have a major concern with the pub night that your organization is holding on May 20, 2006. I would strongly suggest that you hire uniformed police officers to work security at the pub.

You have had serious problems occur at these events every year, and despite the numerous warnings from us, you have not done anything about them. This year we would like you to think and plan ahead to avoid these problems.

Having a officer present at the pub will make your event more enjoyable. The cost of hiring an officer is $75.00 an hour. Call and let me know what you think. I look forward to hearing from you.

Sincerely,

Sergeant Betty Buckins

ACTIVITY: WRITING POLICE LETTERS

Write a letter for each of the following scenarios. Use full-block style and two-point punctuation for each.

Responding to a Complaint

A. You are a staff sergeant working at the Metropolitan Police Department. You received a complaint about how one of your officers handled a domestic dispute. The complainant is one of the persons involved in the dispute. You looked into the incident and feel your officer acted appropriately. Write a letter responding to the complainant.

B. You are Staff Sergeant Fellows working at the Metropolitan Police Department. You received a complaint that Constable Legume mishandled a noise complaint. You investigated and found that the situation needs further investigation by the chief. Write a letter to Maria Briglio, the complainant, informing her of your findings.

C. You are a constable working at the Metropolitan Police Department. You received a complaint from a community member accusing you of backing up your cruiser in the parking lot at the local mall and damaging the bumper of her car. She states you were dispatched to a call, and you didn't stop when the accident occurred. You do not recall backing up and damaging the car. Write Lois Bridges a letter defending your position.

Information Letter

A. You are a staff sergeant working at the Metropolitan Police Department. Write a letter to residents within a certain neighbourhood in your community informing them that their neighbourhood will be heavily patrolled over the next few weeks due to the high number of break and enters in their area.

B. You are Chief Johnson of the Anywhere Police Department. You received a letter from Chief Edwards from the Metropolitan Police Department requesting information on your morality squad. Write Chief Edwards a letter providing him with information on how the morality squad works within your department.

C. You are a staff sergeant working at the Metropolitan Police Department. Write a letter to the local college informing the dean of the criminal justice program that the Metropolitan Police Department is looking for new recruits. Some of the criminal justice students might be interested in applying for a policing position.

Confirmation Letter

A. You are a constable working at the Metropolitan Police Department. You have been invited to be part of a provincial committee looking into the regulations surrounding the use of pepper spray. Write a letter confirming your acceptance.

B. You are Chief Edwards working at the Metropolitan Police Department. You have made arrangements with Chief Migwans for four of your officers to visit his department to receive special weapons training. Write Chief Migwans a letter confirming that your four officers will visit Chief Migwans's department for the training during the week of May 15–19, 2006.

C. You are a staff sergeant working at the Metropolitan Police Department. Dean Williams from the community college has invited you to speak to the students in the criminal justice program about constable positions available at the Metropolitan Police Department. Write Dean Williams a letter confirming that you would be happy to speak with the criminal justice students.

Suggestion Letter

A. You are a staff sergeant working at the Metropolitan Police Department. You need to write a letter to staff of a local arena suggesting that they have uniformed officers present at the community night they are holding.

B. You are a staff sergeant working at the Anywhere Police Department. You are a member of the provincial committee looking at the OPTIC system. Your committee meets every second month, but you feel that since the committee is ready to make recommendations about changes to the system it likely only needs to meet once every four months instead. Write Staff Sergeant Rend a letter making this suggestion.

C. You are a staff sergeant working at the Metropolitan Police Department. You have been noticing that the constables you have hired from the local college recently lack report-writing skills. Write the dean of the criminal justice program a letter suggesting the students receive more report writing as part of their workload.

Request Letter

A. You are a staff sergeant working at the Metropolitan Police Department. You have taken on the responsibility of introducing community policing in your department. Your first step is to gather information about what is happening at other police departments around the province. Write a letter to the Town and Country Police Department requesting any information on its implementation of community policing.

B. You are Chief Edwards of the Metropolitan Police Department. You are unhappy with the structure of your morality department. You have heard that Chief Johnson of the Anywhere Police Department has an excellent set-up. Write Chief Johnson a letter requesting information on his morality squad.

C. You are a staff sergeant working at the Metropolitan Police Department. Annually, the policing association organizes a conference to address current issues in policing. Each year, you encourage the constables in your unit to attend. Write the conference organizer, Staff Sergeant Bovington, a letter requesting information on the details of this year's conference.

Chapter 4

Notebooks

Upon successful completion of this chapter, you should be able to

1. recognize the uses of the notebook
2. prepare a notebook for use
3. recognize and apply the do's and don'ts of note taking
4. record daily information in a notebook
5. use proper interviewing skills to acquire information for investigations
6. record detailed information in the notebook
7. use a proper diagram in the notebook

INTRODUCTION

As a police officer, you are required to keep notes as an essential part of your duties. The entries you make in your notebook are used in many different ways:

1. They will be transformed into a written police report.
2. They will be used to pass on information, provide a permanent record, provide a complete picture, and assist in the legal process.
3. Your original notes will form the basis for all future documents and the investigations that follow.
4. Your notes will help you recall important details of occurrences before you go to court to testify.

PREPARING YOUR NOTEBOOK COVER

Before you begin using your notebook, you need to have the proper identifying information on the cover. If the front cover of your notebook contains a preprinted form to be completed, fill it in before you begin to use the notebook. Write your name, rank, badge number, reference number, date of first entry, and date of the last entry on the cover. On some notebooks, the top binding also provides a place for the date of the first entry, date of the last entry, and reference number. The top binding should also be completed. If your notebook does not include a preprinted form, note your name, rank, badge number, and the date the book was started on the front cover. Once your notebook is complete, note the date of the last entry on the front cover as well.

Here is an example of a completed front cover:

Name ___Sarah Bois_____
Rank __constable_____ Badge No. __217_____
Reference No. __Book #4_____
Date of First Entry __April 16, 2006_____
Date of Last Entry __July 18, 2006_____

Here is an example of what the top binding will look like:

Date of First Entry __April 16, 2006_____
Date of Last Entry __July 18, 2006_____

YOUR NOTEBOOK PAGES

The inside pages of your notebook should be set up in a particular way. Again, depending on the notebook you are using, you may or may not have to do some setting up before you make any entries. Some notebooks have pre-numbered pages and a ruled left-hand margin. If your notebook is already set up this way, you are ready to make entries. If the pages of your notebook are not numbered, number them as you begin each new page. If your notebook doesn't have a left-hand margin, draw one in.

Here is an example of how the pages of your notebook should look:

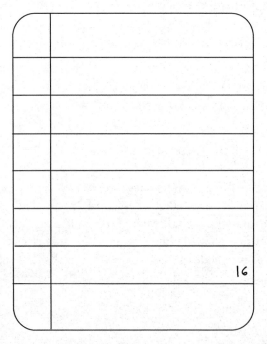

The left-hand margin is needed because it is here that you record the time you begin taking notes. The blank lines next to the time are where you write the information you obtain. Numbering your pages will alleviate any possible future suspicion that pages were torn out.

PRACTICE: PREPARING YOUR NOTEBOOK

Assume you are a constable working at the Metropolitan Police Department. You are ready to begin a new notebook. Prepare your notebook in the proper format (i.e., the cover, top binding, and pages).

TIPS FOR KEEPING YOUR NOTEBOOK – DO'S

There are some things you need to consider when you are using your notebook. The items listed below are rules you should always follow.

- Use only one notebook at a time. When you have finished using all the pages in the notebook, you can move on to the next.
- Use black or blue ink.
- Write neatly and legibly. Doing so will allow readers to understand fully what you have written.
- Write your notes at the earliest opportunity. Don't rely on your memory.
- Keep your sentences short, if possible. This makes your notes easier to read, follow, and understand.
- Keep your notes objective. Avoid writing any subjective material.
- Fill in every line of your notebook. This will avoid suspicions that you previously left spaces and then went back and filled them in. The only blank lines should be a single blank line separating days. The last line at the end of one tour of duty is left blank to separate that day from the next.
- Draw a single line through any corrections. Place your initials beside the line, and then make your correction.
- Record the *who, what, why, when, where,* and *how* in your notebook. (See Chapter Five, pages 98–99 for more information.)
- Record all of your daily tasks in chronological order.
- Use the 24-hour system to record times. This is the most common method used.
- When two or more officers are involved in the same investigation, each officer should take notes in his or her notebook.

TIPS FOR KEEPING YOUR NOTEBOOK – DON'T'S

These items are things you need to avoid when using your notebook.

- **Do not** tear pages out of the notebook. Doing so may create suspicions among defence lawyers and may lead to case dismissals.
- **Do not** use abbreviations in your notebook that are unclear or might be misunderstood.

- **Do not** use slang in your notebook. Again, these expressions may be unclear or misunderstood.
- **Do not** use the notebook for personal reasons. Record only information that deals with your professional obligations.
- **Do not** use the back cover to record any information.

ENTERING DAILY INFORMATION

Each day when you begin your tour of duty, record the following information in your notebook:

- The date — this includes the day of the week, date, month, and year
- The time you have reported for duty
- The weather conditions
- The patrol district assigned
- Relevant messages — important notices/announcements such as a car to watch for

As you work your shift, your notebook should be used to record the following information:

- All the actions you took throughout your tour of duty. It is important to indicate as well the time that these actions occurred.
- All the specific details surrounding an occurrence. (Again, remember the *who*, *what*, *where*, *why*, and *how*.)
- All the statements made by any accused, victim, or witness.

When you have completed your shift, record the following in your notebook:

- The time you reported off duty
- Your rank, signature, and badge number
- Your supervisor's signature follows your signature
- Remember to leave one space blank and before making entries for the next shift.

Here is an example of how one officer, Constable Ekat, kept her notebook during a shift:

Monday, October 3, 2006

0700

Clear, cool temperatures

Assigned district 25B

*Note: Be on the lookout for a Ford 1/2 ton, silver grey, 1997, licence YME 800

0730 -Began patrolling area

-Drove around area

0830 -Received dispatch on possible hit and run at Metropolitan Mall

0845 -Arrived at Metropolitan Mall - west side

-Observed Chrysler Caravan that was damaged on passenger's side

-Passenger's door and front end of bumper were dented in

-Appeared that a car ran into the van

-Talked to May Lynn Prum—owner of the van

-Ms. Prum indicated that she parked her van in the mall parking lot at 0800

-She works at Mel's Investments across the street and often parks at the mall

-She came to her car at 0825 to do some errands for the office and

discovered her car smashed up

-Ms. Prum indicated that she had no idea who did the damage

-Ms. Prum reported that when she came to her car, no one was around

-Ms. Prum's insurance company is Royal Gold—Constable Ekat told her to notify her

insurance company

Ms. Prum—125 Nomis Avenue

946-1452

D.O.B. - May 15, 1963

0930 -Left scene and began patrolling area

0945 -Stopped speeding white Chevy Blazer

-Vehicle was travelling 160 kph in a 80 kph zone

-Owner produced his driver's licence

-Owner of vehicle was identified as Mario Martin.

-Constable Ekat asked Mr. Martin to produce his insurance but he was unable

to locate it

-Constable Ekat informed him that he would have 24 hrs to produce the

insurance

-otherwise he would be given an additional fine

-Mr. Martin stated that he was speeding because he was late for an appointment

-Wrote Mr. Martin a speeding ticket

Mario Martin—138 Blure Street

236-9147

D.O.B - January 4, 1970

1015-Left scene and began patrolling area

1030-Stopped for a break

1045-Began patrolling

1200-Lunch Break

1300-Began patrolling

1315-Dispatched to Parkview Public School on a report of a physical altercation

1320-Interviewed Sam Blinders—grade 8 student, age 13, and

-Miguel Longo, age 13, grade eight

-Principal Francois Bourgeois indicated the two boys were found in a physical altercation. Sam Blinders was found with a knife

-Sam Blinders stated that they had been enemies for the year and he had enough of Miguel

-Sam indicated that he was afraid of Miguel and that is why he had been carrying a knife

-Miguel often threatened Sam by saying "he would be sorry"

-Sam stated the fight broke out when Miguel started teasing Sam about his low math mark

-Sam threw the first punch and then brought out the knife

-Miguel was then interviewed, and he indicated that he and Sam have had a feud going all year, and he was teasing Sam about his low grades

-Miguel indicated that Sam started the fight with a punch, and he became frightened when he saw the knife

-The two students were informed that they were expelled from school for two weeks and their parents were notified

-Constable Ekat took the knife from Sam

1430-Talked to the boys' parents about the occurrence

Sam Blinders	Miguel Longo
12 Simpson Avenue	104 Maple Avenue
784-2680	784-9420
D.O.B. June 3, 1993	D.O.B. Feb.10, 1993

1445-Break

1500 -Patrolled area

1900 -Off duty

Constable Erin Ekat

Badge # 029605

Staff Sergeant Bienvue

PRACTICE: TAKING NOTES

In the notebook that you have properly prepared, make entries for two school weeks. Record your activities for each day, beginning with your arrival at school and ending with your departure. Follow the tips given under "Your Notebook Pages" and the "Do's" and "Don't's" sections.

RECORDING DETAILED INFORMATION

The information you record in your notebook needs to be complete and accurate. Your descriptions of such things as people, property, and cars should be as detailed as possible to make it easy for identification later. Following the details listed below will ensure that you have recorded the information as accurately and as completely as possible.

People

Depending on the situation, you will need to record certain information about a person. For example, if you are taking notes from a victim of a break and enter, you will not need to record details such as the victim's weight, height, and scars, because they aren't relevant. However, if you find yourself taking a description of a possible suspect, these kinds of details would be essential for the investigation and identification. Or, if you have arrested someone, the following details would be necessary:

- name, age, address, phone number
- marital status
- social insurance number, driver's licence number
- place of employment
- person's gender
- race, height, weight, build (slender, medium, heavy)
- hair colour, eye colour
- scars, tattoos, outstanding features

Vehicles

If you are recording information about a stolen car, or a vehicle possibly involved in a hit and run, record the following details:

- owner's name, licence number/year/province
- vehicle year/make/model/colour
- description of vehicle contents (if necessary)

Bicycles

If you are taking notes on a missing bicycle, record the following details:

- owner's name and address
- make/model/colour
- special equipment (light, basket)
- value
- identifying marks

Jewellery

If you are taking notes on reported missing jewellery, record the following details:

- type of article
- date and place of purchase/value at purchase
- size
- engravings or identifying marks

Furniture or Electrical Equipment

If you need to describe furniture or electrical equipment, follow the details listed below:

- type of article/style/age/model/brand name
- colour/material
- value
- identifying marks

Weapons

Use the following details to describe weapons:

- type of weapon (gun, knife)
- model/serial number
- distinctive features
- origin (stolen, purchased, homemade)

Road Conditions

If you need to describe the road conditions, follow the details listed below:

- weather conditions affecting the road
- surface type
- width of road, curbs, and ditches
- hazards
- marks on the road

PRACTICE: RECORDING DETAILED INFORMATION

Following the tips given, fully describe a piece of jewellery, a car, a person, and a TV. Describe them in full detail, as you would for a police report.

INTERVIEWING SKILLS

Interviewing is a technique used by police officers to gather information that could be critical in apprehending and obtaining criminal convictions. An interview involves questioning a victim, suspect, or witness to gather information regarding an actual or suspected crime. When you, as an officer, interview individuals to gather information about an incident, you must be sensitive to the individual's needs, approaching the situation with both tact and determination. You want to be able to gather as much information as possible about the incident. Here are some tips to follow when interviewing someone to ensure that you get everything you want out of the interview:

1. Ask the individual what happened. As this point, do not take any notes in your notebook. Let the individual tell you everything that happened and everything he or she knows. As the individual is telling the story, you should give him or her signals that you are listening and that you care about what happened. You want to build rapport with the victim so that he or she feels comfortable enough to tell you everything he or she knows. It is important that during this step you not take any notes in your notebook and that you simply listen to what the individual has to tell you.

2. Next, guide the individual through the incident. This time, take notes in your notebook. Have the individual retell the story. You should ask questions, guiding the individual through it a second time. While you are taking notes, make sure that you get all the important information that you will need to complete your report. Remember to answer *who*, *what*, *when*, *where*, *why*, and *how*. Also, do not include any information in your notebook that would not be relevant to the occurrence. You will often use your notebook for court; also, defence and Crown attorneys will use it, so make your notes complete and professional.

3. Read your notes to the individual. This step will verify that what you have written in your notebook is accurate. The notes you have taken are what you use to write your narrative. You cannot rely on your memory. During this step, the individual may remember additional information that you can add, or the individual may make some corrections to what you have written.

4. Once you have completed your notes, you may want to complete a *will say*. A will say may be written by the officer and signed by the individual, or it may be written and signed by the individual. A will say verifies what an individual is willing to say and testify to in court.

5. If there is more than one person at the scene whom you need to interview, interview one individual at a time.

6. When you approach a scene and there are many people around where the incident occurred, approach the most obvious individual first; this normally is the person who is talkative and excited.

7. When you are asking questions to gather information, avoid asking, "Did you see what happened?" This may elicit a one-word answer such as "No." You want more than this as an answer.

8. Ask open-ended questions that require more than a one-word response. These types of questions may also lead to other witnesses. "What happened here?" is an example of an open-ended question.

9. Always offer supportive comments if an individual is uncomfortable with relating facts about the incident.

10. Keep the lines of communication open with the individuals whom you interview. Let them know how they can reach you if they remember anything else or if anything else happens.

PRACTICE: INTERVIEWING SKILLS

Choose a partner within the class. Practise your interviewing skills by following the guidelines provided. The individual being interviewed may want to tell the interviewer of the challenges faced in the first few weeks of college. As the interviewer, remember to listen first, then guide the interviewee through the story a second time. Ask open-ended questions to get as much information as possible. Change roles after the first interview is finished.

USING DIAGRAMS IN YOUR NOTEBOOK

Many police officers draw diagrams in their notebooks for particular occurrences. The diagrams may include measurements taken after a traffic accident, the position of a body in a homicide case, the relationship of rooms and objects in a burglary, the location of an injury-causing accident, and the location where contraband was found. Many officers use diagrams in their notebooks for their own purposes. They draw diagrams to help them with recall in court. Many officers also use diagrams to further illustrate unusual circumstances. For example, a suspect may have gained entry into a home in an unusual way. The officer may draw a diagram to show how the suspect gained entry. If an officer feels it is an important piece of information or incident, he or she may transfer the diagram from the notebook into the written police report. If you, as a police officer, decide to use a diagram in your notebook, follow these guidelines:

- Your first notebook diagram should not be altered, since this might cast doubt later.
- Practise a consistent method of drawing diagrams. Of course, the more experience you have as a police officer, the easier it will become.
- Be sure to use fixed objects from which to relate position. For example, a body should be shown in relation to distance from a wall, corner, or doorway as opposed to a table or chair, which can easily be moved.
- Moveable objects may be shown; however, measurements should be taken from something that does not move.
- In traffic cases, lanes should be indicated as well as the distance from the nearest intersection, point of impact, position of the car when the investigator arrived, and any other details from the scene that can be reconstructed.
- If you use symbols in your diagrams to indicate different objects, include a legend explaining what your symbols mean. This way, when you go to court and are reviewing your notes, you can refer to the legend and refresh your memory as to the symbols' meanings.
- Be as clear as you possibly can with your drawings. Others, such as defence lawyers, will see and interpret what you have drawn.
- Be complete. Don't leave anything out of the drawing that may be necessary to the incident and case.

PRACTICE: DRAWING DIAGRAMS

Assume you are a police officer working at the Metropolitan Police Department. You have been dispatched to a car accident. Invent details surrounding a car accident and draw a diagram showing how the impact happened. Use the guidelines discussed to produce your diagram.

DISCUSSION QUESTIONS

1. Discuss the uses of the notebook.
2. List the do's and don't's for recording information in the notebook.
3. Discuss what information should be recorded in your notebook.
4. List some guidelines you should use when interviewing.
5. List some guidelines you should follow when making a diagram in your notebook.

ACTIVITY: RECORDING IN A NOTEBOOK

Choose a partner within the class. Take turns being the individual reporting an incident and the constable taking notes. Each person should choose a different scenario. The person who takes the role of the constable is to interview the individual and take notes in his or her notebook in preparation for completing a police report. The notebook should be properly prepared. Use the tips on interviewing skills, do's and don't's, and diagrams to complete your notes.

Scenario One

You are Juliet Hall and you have called the police to report an assault. You were outside in your yard when you suddenly heard a woman screaming. You turned around and discovered that a young woman and a man were arguing in a driveway two houses away from yours. The young man stood very close to the woman and pointed his finger in her face while yelling. The woman argued, but seemed to be afraid as she was backing up. You then saw the man punch the woman in the face and kick her down. You were very upset at this and ran into the house to call the police.

Scenario Two

You are Sally Smithers. You called the police to report a hit and run. You were outside doing some gardening. Your golden retriever, Sam, was outside with you. You suddenly heard a loud yelp from your dog. You turned around and discovered a car had hit your dog. The driver of the car did not stop and sped away. You noticed the driver was weaving in and out of the lanes. You ran into the house to call the police.

Scenario Three

You are Eugene Mallochio, and you have called the police to report a drunken disturbance. It was early evening (1800) and your two children were outside playing soccer with four other neighbourhood children. Your daughter, age 9, came in upset. She said the neighbour, Sal Nasso, was bothering them. He kept interrupting the game by joining in and falling everywhere. He used foul language around the children. This is not an uncommon occurrence. You know Sal has a drinking problem and has bothered neighbours before. However, this was the first time that he confronted the neighbourhood children. This upset you. You called the children into the house and contacted the police.

Scenario Four

You are George Bingels, and you called the police to report an armed robbery. You were working the night shift at Variety Plus when a masked man entered the store and approached the cash register. The man pointed a gun at your head and told you to open the register. He then told you to lie down on the floor, face down. He took the money out of the register and told you that he would be back for you if you called the police. You called the police when he left.

Scenario Five

You are Aaron Szczepaniak, and you called the police to report a suspicious parked car. You live down the street from a neighbourhood school, and every day you watch the children walk home. Over the last week, you have noticed a car parked across the street, and the driver of the car has been watching the children walk home. The driver, a man, has not picked up any children, but he arrives at the same time each day and leaves once the school rush is over. You are curious about this driver's intentions and decide to report the incident before something happens.

Scenario Six

You are Filomena Carriatore, and you called the police to report a stolen lawn mower and bicycle. When you were about to leave the house one morning, you discovered that you had left your garage door open all night. You usually ensure that you have the garage door down every night before you go to bed, but you must have forgotten the previous evening. You notice that your new Craftsman Lawn Mower and CCM blue 10-speed bicycle have been stolen from the garage. Nothing else seems to be missing.

Chapter 5

Police Report Writing

LEARNING OUTCOMES

Upon successful completion of this chapter, you should be able to

1. define the police report-writing process
2. define how police reports are used
3. recognize and apply the elements of a well-written report
4. define and use first- and third-person voice in police reports
5. use the active voice in police reports
6. write police reports objectively
7. use the five Ws in police reports
8. recognize and use the report-writing process
9. use chronological order in police reports
10. write police report narratives
11. recognize the general occurrence, supplemental, and arrest reports
12. write general occurrence, supplemental, and arrest reports

TWO-PART POLICE REPORTS

A complete police report usually consists of two features. The first part consists of the forms you need to fill out that pertain to the incident to which you were dispatched, if your department is part of the Ontario Police Technology Information Cooperative (OPTIC). These include forms such as

- homicide/sudden death reports
- fraudulent document reports
- observation reports
- guns
- security
- drugs
- miscellaneous articles
- missing-person reports

- vehicle details
- arrest reports
- general occurrence reports
- persons' physical Descriptors
- supplementary reports
- person details
- charge/court details

On the forms, you simply fill in the blanks or check off appropriate boxes that deal with the specific situation. The general occurrence, supplemental, and arrest reports are discussed in more detail in this chapter since they are the most commonly used.

The second part calls for you to write a narrative that describes in detail what happened; what you were told by the witnesses, victims, and suspects about the incident; and what investigation, if any, that you undertook. The narrative is the more difficult of the two to complete because it's where you put your words into action. The entries you collect in your notebook are written into a narrative, and the narrative is primarily what will be of interest to your staff sergeant and the lawyers. Your narrative must be well written, because any little flaws may give the defence lawyer an opportunity to have the court dismiss the charges against his or her client.

Most of this chapter will deal with writing an effective narrative of the police report.

REPORT-WRITING PROCESS

The report-writing process may differ depending on the police department. Here is how the process works:

1. The constable is dispatched to a call.
2. The constable gets to the scene and takes notes in his or her notebook based on what he or she is told and observes.
3. If the police department where the constable works uses a dictation system, the constable will use notebook entries to dictate the police report. A typist will transcribe the constable's dictation and type the report.
4. If the department does not use the dictation system, the constable will write out his or her own report.
5. The completed report is sent to the staff sergeant for review. If the report is incomplete, it will be sent back to the constable for revisions.
6. Once the report is complete and checked by the staff sergeant, it is either filed away if it is not needed for legal purposes, or it is sent to defence and Crown attorneys for review. The defence and Crown attorneys will use the report to help build their cases.
7. Once a case goes to court, the constable will review his or her notebook and police report to recall the incident.
8. The report is filed away for future reference.

REPORT-WRITING PROCESS

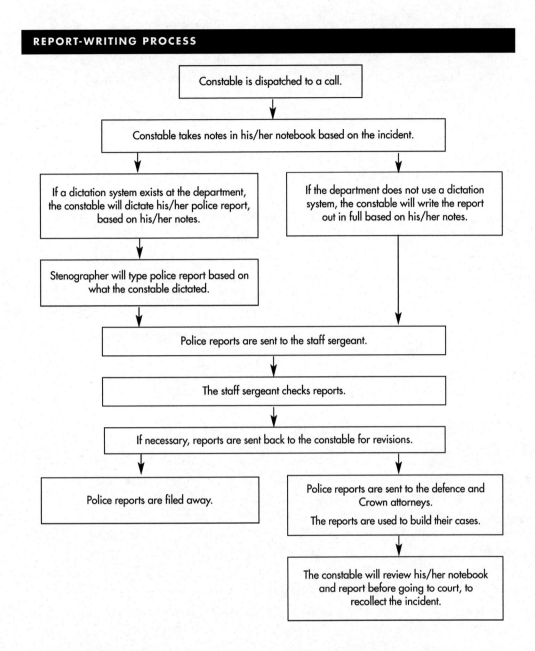

Constable is dispatched to a call.

Constable takes notes in his/her notebook based on the incident.

If a dictation system exists at the department, the constable will dictate his/her police report, based on his/her notes.

If the department does not use a dictation system, the constable will write the report out in full based on his/her notes.

Stenographer will type police report based on what the constable dictated.

Police reports are sent to the staff sergeant.

The staff sergeant checks reports.

If necessary, reports are sent back to the constable for revisions.

Police reports are filed away.

Police reports are sent to the defence and Crown attorneys.
The reports are used to build their cases.

The constable will review his/her notebook and report before going to court, to recollect the incident.

HOW POLICE REPORTS ARE USED

There are many uses for the police report, not only for the police department itself, but also for individuals and organizations outside the department. Let's look at the uses of the report to the department itself.

1. The officer who wrote the report will want to review it before he or she testifies in court.

2. Another officer in the department will want to review the report if this case is turned over to him or her. The officer will want to become familiar with what has happened so far in the case.

3. The staff sergeant is always interested in reading the reports for the purpose of crime analysis. If the staff sergeant notices a crime-related trend, he or she will take steps to ensure that something is done to rectify the problem. For example, if the staff sergeant notices that several break and enters have occurred in a particular area, he or

she may take steps to ensure that the area is heavily patrolled to protect the people in that area.

4. Staff sergeants and other management personnel who evaluate officers' performances may use police reports.

5. The reports act as a permanent record of an officer's actions and provide evidence of the police department's responsiveness to the community and its needs.

6. The reports provide the police department with statistical information to confirm the need for more officers or for special projects that they need to work on or promote.

Individuals and agencies outside the department may use police reports.

1. Both defence and Crown attorneys use the reports to build their cases for court.

2. Police departments across Ontario provide Statistics Canada with information gathered from police reports for statistical purposes. An example of this is Statistics Canada's calculation of the number of break and enters in Ontario/Canada during the year.

3. Insurance companies are interested in police reports to determine the validity of claims and for statistical purposes.

4. The media use police reports to relay news of interest to the community.

CHARACTERISTICS OF A WELL-WRITTEN REPORT

As in any other type of writing, police reports must be written well. Remember, when you are preparing a police report, your writing should display professional qualities. Many people are going to read your reports, so you must ensure that the paperwork you present contains the information they will need. You must also present your information professionally, in a way that can be used to convict a suspect. To achieve this, your report should be

1. **Accurate** — It is crucial that all the information you include in your report be accurate. Remember that many of your reports are used in court. Make sure that you have recorded all names correctly, as well as the facts of the incident.

2. **Complete** — Answer the *who, what, when, where,* and *how* in your report. The *why* is sometimes difficult to answer, so do not include this in your report unless you know with certainty why something happened the way it did.

3. **Clear** — Use language that is easy to understand. Write so there is no confusion. Again, keep in mind that your reports may be used in court. A confusing report increases the chances of a suspect's being set free.

4. **Concise** — Get right to the point. Tell the reader what he or she needs to know about the incident and move on. The reader will be interested only in important information that pertains to the case.

5. **Objective** — Keep all your prejudices and opinions out of your report. Report the facts of the incident only. This is important because it can determine the outcome of a case.

6. **Factual** — Record information that is verifiable. This includes things that are observable.

7. **Written in the active voice** — Writing in the active voice enables the reader to see immediately what the action was and who performed it.

8. **Written in the first-third-person point of view (POV)** — Depending on the police department, you may write your report in either the first or third person. Whatever point of view you use, keep it consistent throughout the report.

9. **Written on time** — People need to read your report, so do not procrastinate. Both defence and Crown attorneys will want your report as soon as possible. They have cases to prepare.

THIRD PERSON VERSUS FIRST PERSON

Although the majority of police departments and officers prefer to use the third person POV in report writing, much depends on the police department. If the department wants you to use the first person, this means that in your reports you should refer to yourself as *I*. If your department prefers the third person, write your reports using your title and last name. For example, if your name is Joe Scott, you would refer to yourself as Constable Scott in your report. You would not say, "I arrived at the scene of the accident at 1700." You would say, "Constable Scott arrived at the scene of the accident at 1700."

Here are some examples of first-person and third-person statements in police reports:

First Person	Third Person
I arrested the accused for drinking and driving.	Constable Bulldry arrested the accused for drinking and driving.
Constable Kwan and I were dispatched to a break-and-enter call.	Constable Kwan and Constable Sanders were dispatched to a break-and-enter call.
Constable Nixon called me in to investigate further.	Constable Nixon called Sergeant Barry in to investigate further.

PRACTICE: THIRD-PERSON WRITING

This police report was written in the first person. Rewrite the report making it a third-person narrative.

On Monday, July 3, 2006, at 2330 Constable Jackson and I were dispatched to a report of a physical altercation at Reggie's Saloon. When we arrived at the bar, two men were fighting in the middle of the dance floor. I separated the two men, and Constable Jackson and I interviewed them.

Sam Stevens told me the following: He and Irvin Innes work together at RJ Products, and they were told today at work that they would be receiving raises. To celebrate, Stevens and Innes decided to go and have a few drinks after work. Stevens stated that the two arrived at the bar at 1730 and had been drinking since. I asked Stevens how the fight started, and he said that Innes was jealous of his work at the factory and angry that Stevens made more money than he did. Stevens told Innes how much of a raise he was getting, and Innes became angry. He started calling Stevens names and saying he wasn't worth the money he was being paid. I asked Stevens who threw the first punch, and Stevens responded that he had.

I then interviewed Innes, who stated that Stevens started the fight. He stated that Stevens has a bad temper and can't take a joke. He said he was just joking with him about his work at the factory.

At this point, I told both men that they would have to leave the bar. I questioned the men as to how they got to the bar and they both responded that they drove their cars. I informed the men that they could not drive and that we would drive them home.

I informed each man that he was to stay clear of Reggie's Saloon for a while or next time they would be placed in jail for the night. We dropped each man off at his residence.

WRITING IN THE ACTIVE VOICE

When you are writing your police reports, you should use the active voice. Using the active voice helps your reader to identify quickly what the action was and who performed it. Writing in the active voice also makes your writing more concise and reduces the number of words you use. Again, you need to consider that you are writing a police report. Get right to the point and tell the reader what he or she needs to know. The opposite of the active voice is the passive voice. Writing in the passive voice can make it more difficult for the reader to determine who performed the action. Here is an example:

The inmate was stabbed with the knife.

This sentence is written in the passive voice. It is difficult for the reader to determine who performed the action because the actor is not stated.

Inmate Judson stabbed inmate Roberts with a knife.

This sentence is written in the active voice. It is clear who performed the action and what the action is.

To write in the active voice, follow these steps:

1. Identify the action of the sentence.

2. Identify the subject of the sentence, that is, the doer of the action.

3. Write the sentence, ensuring that you place the doer of the action immediately before the action.

Here is an example:

Action: stabbing
Doer: Inmate Judson
Sentence: Inmate Judson stabbed inmate Roberts with a knife.

PRACTICE: ACTIVE VOICE

Identify whether each of the following sentences is written in the active or passive voice. If a sentence is written in the passive voice, change it to the active.

1. The suspect was read his rights.

_____

_____

2. Sergeant Millward warned the constables that their behaviour would not be tolerated.

_____

_____

3. The training session was being presented by the training department of the police services.

_____

_____

4. The meeting was called by Sergeant Nichols.

_____

_____

5. Constable Bourgeois found the weapon on the floor.

_____

_____

6. The report was prepared by the typist.

_____

_____

7. The contraband was hidden under the inmate's bunk.

_____

_____

8. The constables were taking part in the community events.

_____

_____

9. The constable was injured by a gunshot wound.

_____

_____

10. The inmates were told to go to their cells immediately.

_____

_____

WRITING OBJECTIVELY

Writing objectively means that you do not let emotion or prejudices influence your writing. You record only the facts. This is very important in police report writing. Make sure that what you record is verifiable. For example, in your report you should not write "The suspect was drunk." You would need to verify that statement. Instead, you should write, "The suspect appeared to be drunk. He was unsteady on his feet, slurred his words, and was staggering." You must be able to justify your comments. Otherwise, they may be interpreted as subjective.

To write objectively, consider the following:

1. Do not use slang.

2. Do not relate only one side of the story; tell all sides of it. If you fail to do so, you will be accused of slanting the story and favouring one side over the other.

3. Record information that is relevant to the situation and ignore those aspects that have no bearing on it. For example, if you are interviewing a neighbour about what he or she saw during a break and enter, it is not important to record in your report that the neighbour dislikes the victims. This is not relevant to the case.

4. Leave your personal feelings and prejudices out of the report. If you felt that a witness had been overly intrusive, you would not say that in the report.

5. Record only the facts. This means that you should record only the information that is verifiable through your senses (seeing, hearing, touching, tasting, or smelling). You would not write "The victim was suicidal." You would write "The victim appeared to be suicidal. She was unresponsive to questioning, was crying, wanted to be left alone, and kept saying she wanted to die."

6. If you record your opinion in the report, state it as an opinion. Make the reader aware that it is an opinion.

7. Use denotative words in your report as opposed to connotative words. Denotative words are non-emotional and explicit. Connotative words suggest or imply something beyond their literal meaning.

PRACTICE: OBJECTIVE WRITING

Rewrite this police report, making it as objective as possible by eliminating all slang, slanting, spelling errors, connotations, and subjective comments. Also, change the passive voice to the active voice and change the report to the third-person form.

At 1020 hours on Friday, March 11, 2006, Constable Bob Brown and I were on routine patrol when we noticed the accused cruising west on Northern Avenue at Willow in a rust-covered Ford, Licence OFO 312.

Hanging from the mirror was fuzzy dice, and the accused was listening to Def Leppard. The accused's vehicle was weaving from the north lane to the south lane on several occasions. The accused's actions were observed by Constables Bob Brown and me. The accused was asked to pull over by me in front of residential number 280 Northern Avenue. The accused appeared to be drunk. When the accused spoke his breath reeked and his words were pronounced indistinctly. The accused was asked to produce a driver's licence and handed me a Social Insurance Card. The accused was identified at this point as Marty Stafford. I believe he was involved in the break and enter last week at Jack's Video on Queen Street. The accused was asked to step out of his broken-down truck at which time he was observed to be unsteady on his feet.

The accused was arrested by me, and I read him his rights. It was also noticed at this time that the accused has a tattoo on his left arm. The accused was transported to the station where he supplied samples of his breath to the Breathalyzer technician, with the results of 130 and 140 mg.

WRITING THE NARRATIVE

Once you have taken notes in your notebook, your next step is to make your notes into a narrative. To be complete, your narrative needs to contain the *who*, *what*, *when*, *where*, *how*, and *why*.

Who?

Answer specifically who was involved. Include all victims, witnesses, and suspects.

What?

Try to determine what took place. Was it a crime? Was it a civil problem? What did the suspect take? What weapon did the accused use? What damages occurred?

When?

What time did the incident occur? When did you get to the scene? What time did you leave the scene?

Where?

Where did the incident take place? Where did the suspect gain entry?

How?

Try to answer the questions that would explain how the situation occurred. For example, explain how the accused threatened the victim. How did the accused harm the victim? How did the accused gain entry? How did the accused leave?

Why?

This question is often difficult to answer, and unless you specifically know the answer to the *why*, you should not answer it. If you are unsure, leave it out. If you know why the situation occurred, include the reason in your report.

If you are just starting your career as a police officer, you may want to use an outline until you become good at writing police reports. Using an outline will help to make your reports complete and ensure they are written in an organized fashion that is easy to follow. An example of an outline format is provided on page 102.

Take the outline and either dictate your report to the typist or write it in full. Either way, you need to ensure that you have an organized, well-structured report. As with any type of writing, you should have an introduction, a body, and a conclusion.

Introduction

In your introductory paragraph, you need to tell the reader who you are, the kind of incident to which you have been dispatched, where you've been dispatched (address/location), and the time of the dispatch.

Body

In the body of the report you give all of your witness, victim, and suspect accounts. To differentiate among these individuals, use a new paragraph to describe what each person told you.

Conclusion

In your concluding paragraph, tell the reader what happened as a result of your taking the report. For example, if you took a report on a break and enter, you may have advised the victim to contact his insurance company. You may also have told the victim to contact you if he or she remembered any additional information. To close off your report, tell the reader what time you left the scene.

Editing and Rewriting

1. Once you have written the first draft of your report, look it over and edit your work. Ask yourself the following questions to see whether or not you have written your report in the proper format:

   • Have I written my report in the active voice?

- Have I written my report in an objective manner?
- Have I used the third-person or first-person voice consistently in my report?
- Have I answered the *who, what, when, where, why,* and *how* in my report?
- Have I provided the necessary information in the introduction, body, and conclusion of my report?
- Is all the information I provided in my report accurate?
- Is my report free of grammatical errors?
- Is my report written in chronological or spatial order?
- Is my report cohesive? Do its parts fit together nicely, and is there smooth transition from one idea to the next?

2. Once you have edited your work, your next step is to rewrite your report, making all necessary changes. Rewrite and edit until you are satisfied with a final version.

PRACTICE: NOTE-TAKING

Choose a partner within the class. Take turns being the victim and the constable taking notes at the incident. Each person is to complete a different scenario. The person who takes the role of the victim is to tell the constable what happened, and the constable is to take notes in preparation for completing a police report. Remember to use the note-taking process discussed. The victim is to choose from one of the four scenarios provided below. After you have completed your notes, write the narrative for the police report.

Scenario One

You are Olivia Bertolo and you have called the police to report a domestic dispute. Your husband, Reginald Bertolo, has physically assaulted you. He punched you in the face and arm, leaving you with bruises around your eye and on your left arm. He also pulled your hair and kicked you in the ribs. You believe your ribs are broken. Your shirt and skirt have been ripped. The dispute happened when you approached your husband about his drinking problem. He became angry with you and began the assault. After he assaulted you, he left the house. You want the police to arrest him and charge him with assault.

Scenario Two

You are Miriam Ceesay, and you call the police to report a stalking. Your 13-year-old daughter, Debra, has been feeling frightened and upset lately. A neighbour, Jack Mauthine, has been following her around, calling her on the phone late at night, and writing her letters. Jack is 20 years old and is now beginning to frighten Debra. You have repeatedly asked Jack to discontinue his behaviour but he has not, despite your requests. This behaviour started with phone calls. During his phone calls, Jack repeatedly told Debra that she was beautiful and that she had a wonderful figure. He also told her she should not date any of the boys at school, but that an older guy like him would be better for her. He then began writing her letters. His letters have been much the same. He states over and over again how beautiful he thinks she is and that she deserves a guy like him. Recently, Jack has started following Debra around. The other day she noticed that Jack was walking about 30 feet behind her, following her home from school. Debra and her friends also noticed Jack behind them at the mall the other day. Your daughter has become frightened and is afraid to leave the house. You want the police to put a stop to Jack Mauthine's behaviour.

Scenario Three

You are Karen Vesters, and you have called the police to report a hit-and-run accident. You were at the mall in the parking lot, waiting your turn to back out of your parking spot to leave. A car backing out of its spot hit your car, an Oldsmobile Cutlass, in the rear bumper. The impact was hard enough that your bumper and rear end were dented. The trunk of your car no longer closes because it was also smashed. The driver of the vehicle who hit you did not stop but immediately raced off. His car was also damaged. You managed to get a glimpse of the first three digits of his licence number, which were 303. His car was a grey Buick Le Sabre.

Scenario Four

You are Mark Machini, and you call the police to report stolen property. You believe that your cottage neighbour, Ralph Reed, has taken your firewood from your yard. Ralph uses his cottage year-round, but you use yours only during the summer months. Last autumn, you purchased and chopped four cords of wood for the next spring. Ralph came over to your cottage when you were chopping the wood and commented that he had priced wood, but it was too expensive. He said that, as a result, he wasn't going to buy wood and that he wouldn't be able to use his cottage throughout the winter. When you returned to your cottage in the spring, half your wood was gone. You suspect it was Ralph. You know he managed to use his cottage throughout the winter. You report this incident to the police.

CHRONOLOGICAL ORDER

When you are writing your police reports, you need to arrange your ideas in a way that is easy for your reader to follow. In your report, you cannot jump from one event to another or from a victim to a witness. This will confuse your reader. You need to ensure that your report is easy to follow, making it clear to your reader what happened when and who told you what. You can do this by using chronological order. Chronological order can be defined as describing the events as they happened in time. You are taking a report on a break and enter, and the victim tells you that he arrived home, noticed his door was open, went inside and saw that his belongings were thrown all around his house, and then noticed that his TV, VCR, stereo, paintings, and jewellery were stolen. You, as the reporting officer, would want to record the events in the order in which they happened or were discovered by the victim. Do not jump from one thing to another. In other words, in your report, do not write that the victim had property stolen and then say that when he got home the victim discovered his door was open. This is confusing and hard to follow. It may also leave questions for the defence lawyer in court. Tell the story in the order that the events occurred. In the case of the break and enter, you would describe the occurrence starting from the time the victim got home and discovered the break and enter, and ending at the time you left the scene.

Also, describe in your report what you were told by the victims and the witnesses, one at a time. Do not jump from one person to another in the same paragraph. Let's look at the break-and-enter example. If a neighbour saw something relevant to the case, you would report what the neighbour saw in a paragraph separate from that describing what the victim saw. Otherwise, your report might be confusing. Keep each person you interview separate. Also, arrange them in the order in which you interviewed them.

USING AN OUTLINE FOR REPORTS

Introduction

Who you are: _____

What kind of incident you were dispatched to: _____

Date, time, and place you were dispatched: _____

_____

Body

What you were told by victims, witnesses, suspects:

Account #1. Name/address/phone number of person you interviewed: _____

_____

_____

What did this person tell you? _____

_____

_____

_____

Account #2. Name/address/phone number of person you interviewed: _____

_____

_____

What did this person tell you? _____

_____

_____

_____

Conclusion

What happened as a result of the report? _____

_____

_____

Time you left the scene: _____

PRACTICE: CHRONOLOGICAL ORDER

Place the following items in chronological order. Number each idea. Start by placing the number 1 beside the event that happened first.

_____ Constable Banerjee left the scene at 0230.

_____ Janice Lamothe called over to the neighbours' house at 2300 and asked them to turn down the music because it was difficult for the children to get to sleep with all the noise.

_____ At 0210, Constable Banerjee went over to 234 Bloor St., Apartment #6, and asked the occupants of the apartment to turn down the music and the visitors to vacate the premises.

_____ At 0145, Janice Lamothe decided to call the police after she made two attempts to ask the neighbours to turn down the music.

_____ Constable Banerjee was dispatched to 234 Bloor St., Apartment #5, at 0200 where he spoke with Janice Lamothe, who was complaining of a loud party going on next door to her.

_____ Janice Lamothe stated that at 0100 she went over to her neighbours', knocked on the door, and asked them to please turn the music down. She stated that the neighbours responded "no." At this point, Janice told them she would notify the police.

_____ Janice Lamothe told Constable Banerjee that the neighbours at 234 Bloor St., Apartment #6, had been partying, playing loud music, screaming, and laughing since 2000. Her children had a hard time going to sleep because the noises frightened them.

SAMPLE POLICE REPORT: BEGINNING TO END

Here is an example of the process that a police officer went through to complete her police report:

Constable Simpson was dispatched to a break-and-enter call at 0100 at 853 Someplace Street where she interviewed James Nowell, the owner of the house.

Step One

Constable Simpson asked James Nowell what happened. Constable Simpson listened as the victim told the story the first time. She nodded and gave the victim signals that she was listening.

Step Two

Constable Simpson began to record notes in her notebook by asking the victim questions and guiding him through his story, making sure that he retold his story in chronological order. Here are the notes that Constable Simpson obtained:

- dispatched to a break-and-enter call at 0100 at 853 Someplace Street on June 10, 2006
- interviewed owner of the house, James Nowell
- James Nowell told Constable Simpson the following:
- he arrived home at 0030 and noticed that the side door was open

- he was certain that when he left his home at 2030, he had closed and locked it
- he entered the house and noticed that someone had trashed his kitchen
- dishes had been thrown around and cutlery was all over the floor
- Nowell checked the living room and noticed that the TV, DVD player, and stereo had been stolen
- Nowell described the TV as an RCA 33", the DVD player as a Panasonic, and the stereo as a Pioneer system
- he estimated these three items to be worth $6000
- drawers in the vanity in the main bathroom had been removed and thrown on the floor
- his bedroom closet was emptied, and his clothes were thrown on the floor in that room
- Nowell estimated that the total damage to his house was $5000
- checked the door where the suspect gained entry and it appeared that forced entry was not used
- called the police station and asked for a fingerprint technician to come to the scene of the crime
- asked the victim to call the station if there was anything else he remembered or needed to add
- informed the victim to notify his insurance company regarding the incident
- left scene at 0145

Step Three

Constable Simpson read her notes back to the victim, James Nowell. He confirmed that everything in the report was accurate. Constable Simpson also had the victim complete a *will say* to attach to her report.

Step Four

Constable Simpson went back to the police station and began to transform her notes into a police report by filling in an outline. See page 105.

Step Five

Once Constable Simpson completed her outline, she was ready to write the first draft of the narrative of the report. See page 106.

POLICE REPORT OUTLINE

Introduction

Who you are: Constable Simpson

What you were dispatched to/time you were dispatched/where you were dispatched to: date/break/enter/0100/853 Someplace Street/June 10, 2006

Body

Account #1. Name/address/phone number of person you interviewed: James Nowell – owner of 853 Someplace   /   729-0058

What did this person tell you? – arrived home 0030
– noticed side door was open
– door was closed/locked when he left at 2030
– someone trashed kitchen—dishes/cutlery thrown everywhere
– noticed TV, DVD player, stereo were stolen
  TV – RCA 33"   DVD player – Panasonic   Stereo – Pioneer system
– estimated value $6000
– bathroom vanity drawers emptied thrown everywhere
– bedroom – closets emptied & clothes thrown everywhere
– estimated damage $5000.
– I then checked door – appears no forced entry was used
– called for fingerprint technician to come to the scene of the crime

Conclusion

What happened as result of the report? advised victim to call if remembers anything else or wants to add something
– advised to notify insurance company

Time you left the scene: left at 0145

Constable Simpson was dispatched to a break-and-enter call at 0100 at 853 Someplace street on June 10, 2006.

Constable Simpson interviewed James Nowell, the owner of 853 Someplace Street. His phone number is 729-0058. Mr. Nowell told Constable Simpson the following: He arrived home at 0030 and noticed that his side door was open. He was sure that when he left his residence at 2030 the door was closed and locked. When he entered his pad he noticed that someone had trashed his kitchen. Dishes and cutlery were throne everywhere on the kitchen floor. Next, Mr. Nowell went into the living room where he noticed that the culprit had stolen his RCA 33" TV, Panasonic DVD player, and Pioneer stereo system. Mr. Nowell stated that he bought the merchandise at Leo's Electronics, and that the value of this merchandise was around $6000. Constable Simpson believes this estimate is a bit high. Mr. Nowell then checked his bathroom. The vanity drawers were emptied and thrown everywhere. Also, the bedroom closet was emptied, and his clothes were thrown on the floor.

Constable Simpson then proceeded to check the side door where the suspect gained entry. It appeared that no forced entry had been used. Constable Simpson believes the suspect may have had a key to Mr. Nowell's residence. Constable Simpson then called the station are requested that a fingerprint technician come to the scene of the crime.

Constable Simpson advised the victim to call the station if he remember anything else about the incident. Constable Simpson also advised the victim to notify his insurance company. Constable Simpson left the scene at 0145.

Step Six

Constable Simpson was now ready to edit her work. She looked over her report and made the following changes.

Constable Simpson was dispatched to a break and enter call at 0100 at

853 Someplace Street on June 10, 2006.

Constable Simpson interviewed James Nowell, the owner of 853 Someplace. His phone

number is 729-0058. Mr. Nowell told Constable Simpson the following. He arrived home at

0030 and noticed that his side door was open. He was sure that when he left his

residence at 2030 the door was closed and locked. When he entered his ~~pad~~ *home* he noticed

that someone had trashed his kitchen. Dishes and cutlery were ~~throne~~ *thrown* everywhere on

passive – change to active

the kitchen floor. Next, Mr. Nowell went into the livingroom where he noticed that the suspect

subjective ← <u>culprit</u> stole his RCA 33" TV, Panasonic DVD player, and Pioneer stereo system. Mr. Nowell stated

not necessary ← that ~~he bought the merchandise at Leo's Electronics and~~ the value of this merchandise

this is subjective – take out

was around $6000. Constable Simpson believes this estimate is a bit high. Mr. Nowell then

the contents were

checked his bathroom. The vanity drawers were emptied and thrown everywhere. Also,

the bedroom closet was emptied, and his clothes were thrown on the floor.

Constable Simpson then proceeded to check the side door where the suspect

– subjective – this is an opinion

gained entry. It appeared that no forced entry was used. (Constable Simpson believes the suspect may have a key to Mr. Nowell's residence. Constable Simpson then called

the station ⟨are⟩ *and* requested a fingerprint technician to come to the scene of the crime.

Constable Simpson advised the victim to call the station if he remember ᵉ⁽ˢ⁾ anything

else about the incident. Constable Simpson also advised the victim to notify his

insurance company. Constable Simpson left the scene at 0145.

DISCUSSION QUESTIONS

1. List five types of reports that a police officer would write.
2. List three organizations/agencies other than the police services that would read police reports.
3. List two benefits that police reports might have to outside persons or agencies.
4. List the characteristics of a well-written report.

ACTIVITY: WRITING POLICE REPORTS

Scenario One

You are Constable Lambert working for the Metropolitan Police Services. You were dispatched to 123 Somewhere Street at 0030 to take a report on a break and enter. The following are the notes you took when you interviewed the victim of the incident, Ms. Rose. From the notes, write the narrative of the police report.

Victim — Ms. Andrea Rose
Address — 123 Somewhere Street
Phone number — 765-0932
Date — May 14, 2006

- interviewed Ms. Rose, and she told Constable Lambert the following:
- left her residence at 2030
- residence was properly secured, doors and windows were locked
- bedroom window was not broken
- drawers and furniture were in normal order when she left her house
- returned shortly after 2330 and found that the bedroom window had been forced and the house had been ransacked
- a screwdriver was found on the kitchen table — didn't belong to Ms. Rose
- stolen were: jewellery
    - one carat diamond ring
    - 63.5 cm gold chain (14 carat)
    - sapphire/diamond bracelet
    - diamond earrings (one carat each)
- damage done to the residence
    - broken bedroom window
    - broken dresser drawers
    - kitchen table chairs — broken legs

Interviewed Ms. Juanita Garcia
Address — 125 Somewhere Street

Phone Number — 777-9988
Date — June 3, 2005

- Ms. Garcia told Constable Lambert the following:
- noticed at 2245 that the bedroom window of Ms. Rose was broken and open
- she also noticed at about 2300 that there was a strange car parked outside — she had never seen it in the neighbourhood before
- she then noticed a man running to the car from the Roses' house — she did not recognize the man
- she described the suspect as being about 1.83 m tall, grey, short hair, approximately 91 kg, approximately 45 years old
- suspect also had a noticeable limp — he was limping with his right leg when he ran to the car
- she described car as a Honda – silver in colour
- she did not take down the licence plate number at that time
- Constable Lambert left the scene at 0145
- advised victim to notify her insurance company

Scenario Two

You are Constable Silvaggio working for the Metropolitan Police Services. You were dispatched to 408 Here Street at 2015 to take a report on a missing child. The following are the notes you took when you interviewed Mrs. Martha Butkowski, mother of the missing child. From the notes write the narrative of the police report.

Missing child's name — Sandra Butkowski   Parents' names—Martha and Doug Butkowski
Age — 9 years old   Phone number—555-1010
Birthdate — January 7, 1996
Address — 408 Here Street
City — Somewhere, Ontario
Date — June 3, 2005

- arrived at scene at 2030
- interviewed Martha Butkowski, mother of missing child
- Sandra left the house at approximately 1800 after supper and said she was going to take a bike ride over to Blue Forest Park with her friend Mary Jane Jarvis
- Mrs. Butkowski told her daughter to be home at 1930
- Sandra agreed and left
- at 2000, Sandra still wasn't home, so Mrs. Butkowski drove over to the park to pick her daughter up
- Sandra wasn't there, and neither was Mary Jane Jarvis
- other kids were at the park playing baseball
- Mrs. Butkowski proceeded to Mary Jane Jarvis's house to see if the girls were there
- Mary Jane was home, but Sandra wasn't there
- Mary Jane indicated to Mrs. Butkowski that Sandra had been there, but Mary Jane did not want to go to the park with her, so Sandra went alone
- Mrs. Butkowski went back to the park and asked Joey Mayville (one of the kids playing baseball) whether he had seen Sandra
- he told Mrs. Butkowski that a man who he thought was Sandra's father had lifted Sandra from a swing and put her into his truck
- Joey told Mrs. Butkowski that Sandra was crying, but he assumed she was crying because she did not want to go home
- Mrs. Butkowski stated that the man who picked Sandra up was not her father — Sandra's father is away on business
- at this point, Mrs. Butkowski went home to call the police
- Sandra's description:
    - 1.3 m tall

- approximately 33 kg in weight
- sandy blonde hair, mid-back length, tied in a ponytail
- blue eyes
- wearing:
    - Reebok running shoes
    - Levi blue jeans
    - red Roots sweatshirt
    - jean jacket

- Sandra is a diabetic and depends on her daily insulin shots
- Mrs. Butkowski gives Sandra her insulin shot every morning. Without her shot, Sandra can become very ill
- Sandra was wearing a medic alert bracelet on her left wrist
- Sandra's bike — red CCM 10-speed
Left scene at 2115

Scenario Three

After you finish taking the notes from Mrs. Butkowski, you decide to go and interview Joey Mayville about what he saw happen to Sandra Butkowski at the Blue Forest Park. The following are the notes you took when you interviewed Joey Mayville, a witness to the kidnapping of Sandra Butkowski. From the notes, write the supplemental narrative of the police report.

Witness — Joey Mayville   Parents' names—Molly & Fred Mayville
Address — 72 Bayview Road
Phone number — 968-0031
Birthdate — July 7, 1991
Date — June 4, 2005
City — Somewhere, Ontario

- interviewed Joey Mayville at 2130, and he told me the following:
- He saw Sandra on the swings at the Blue Forest Park, and she was alone
- A man, whom he assumed was her father, came and lifted her off the swings and put her into his truck
- Sandra was crying and struggling, but Joey thought she was crying because she didn't want to leave the park
- Joey described the man as approximately 1.83 m tall
    - moustache
    - dark hair, shoulder-length, curly
    - wearing a blue baseball cap
    - wearing blue jeans and a red sweatshirt
    - truck — red, half-ton
- after he put Sandra in the truck, he picked up her bike and put it into the back of the truck
- after this, he drove off
- Constable Silvaggio informed Joey that if he remembered anything else about the situation that he should have his parents notify the police station immediately
- I left the Mayville residence at 2215

Scenario Four

You are Constable Nguyen, and you are working for the Metropolitan Police Department. You have just been dispatched to a call where a person is believed to have some information about the missing child report. It is 1845 when you are dispatched to the residence of Adele Tosello of 93 Tragina Avenue to take a report. The following are the notes you took from the interview. Write the supplemental narrative of the police report.

Witness — Adele Tosello
Address — 93 Tragina Avenue
City — Somewhere, Ontario
Phone number—621-0412

Birthdate — February 15, 1963
Date — June 5, 2005

- interviewed Adele Tosello at 1855, who claimed to have information about a missing child
- on June 4, she was in her living room watching television and could see directly across the street into her neighbour's driveway
- she saw her neighbour, Bruce Nixon, pull into his driveway at 1935
- she saw him get out of the truck and go over to the passenger's side and carry out a young girl in his arms
- Ms. Tosello at this time was not suspicious because Bruce Nixon does have a niece
- later on in the evening Ms. Tosello also saw Bruce take a bike out of his truck
- it was not until Ms. Tosello saw the report on television on June 5, 2005, that she became suspicious
- when asked if she could describe the child he was carrying, Ms. Tosello answered that it was getting dark out, but she could see that the child had a ponytail
- Constable Nguyen notified the witness to contact the department if there was anything else she remembered
- Constable Nguyen left the witness's residence at 1930
- at 2000, Constable Noble and Constable Nguyen got a search warrant to search the home of Bruce Nixon, address 95 Tragina Avenue
- presented Bruce Nixon with the search warrant at 2015 and told him that we were looking for a missing child
- he responded that the idea of finding a missing child in his house was ridiculous
- Constable Noble and Constable Nguyen searched the house and discovered a cold cellar with a padlock
- Constable Nguyen cut the padlock and found a child who fit Sandra Butkowski's description
- Sandra's arms and legs were tied together, and her mouth was taped
- Sandra was unconscious
- Constable Noble called for an ambulance
- Bruce Nixon was read his rights and arrested
- Sandra was taken to the hospital for observation
- Mr. And Mrs. Butkowski were notified
- Constable Noble and Constable Nguyen then discovered Sandra's bike in the rafters of the garage
- officers left the scene at 2120

Scenario Five

You are Constable Leblanc working for the Metropolitan Police Department. You were dispatched to 461 Hunter Drive at 0100 to take a report on a break and enter. The following are the notes you took when you interviewed the victim of the incident, Dr. Erin Kelly. From the notes, write the narrative of the police report.

Victim — Dr. Erin Kelly
Address — 609 Summer Street (home)
461 Hunter Drive (medical building)
Phone number — 645-0419 (home)
864-4873 (office)
Date — July 7, 2006

- interviewed Dr. Erin Kelly at 0110 and she told Constable Leblanc the following:
- a neighbour living next to her medical building had called her at home at 0040 to say he had heard noises coming from her office building

- neighbour's name is Jack Killorn of 463 Hunter Drive
- Dr. Kelly stated that it was at this point that she thought she had better go check out the situation at her building
- her medical office had been broken into and 10 prescription pads and 3 bottles of morphine tablets (100 tablets in each bottle) had been taken
- prescription drugs were locked in a cabinet that had been smashed open
- prescription pads were located in a drawer in the examination room
- Dr. Kelly stated that earlier that day she saw a patient who was complaining of leg cramps and wanted painkillers
- he was not a regular patient of Dr. Kelly
- Dr. Kelly did not write out a prescription for him because she stated that she saw no medical evidence to prescribe medication
- the patient gave his name to be Pierre Bourdois
- patient became angry when Dr. Kelly would not prescribe a painkiller, and he kept insisting that he needed something to help him cope with the pain
- the patient became aggressive with Dr. Kelly by screaming at her saying, "I need something right now, and you need to give it to me! If you don't give me some medication, you will be sorry"
- Dr. Kelly called security to have the patient escorted out of the building
- Pierre Bourdois gave his health card number as B6278-Yv329
- Constable Leblanc examined the cabinet that had been broken into. It was smashed open, and broken glass was everywhere
- Constable Leblanc examined the back door to the medical building where the suspect gained entry. It appeared as though an axe was used to gain entry
- Constable Leblanc notified Dr. Kelly that further investigation would take place
- Constable Leblanc left the scene at 0215

ACTIVITY: TAKING NOTES AND WRITING POLICE REPORTS

Choose a partner in the class. Four scenarios are given below. Each person is to complete a different report. Take turns being the constable called on the scene to take the report and the victim giving the account. As the constable, take notes on the incident based on what the victim tells you and then complete the police report.

Scenario One: Victim A

You are Stanley Michaud, and you are working at Lester's Convenience Store. You are working the night shift, and at 0230 a man comes into your store, points a gun at you, and demands that you open your cash register. You do as the robber asks, and he takes $300 from your register. He then orders you to open the safe. Again, you comply with his demand, and he takes the $2000 that was in the safe. He then tells you to lie face down on the floor, and he flees from the store. You then phone the police.

You can describe the suspect as

- approximately 1.8 m tall
- approximately 81 kg in weight
- short black hair
- white male
- wearing khaki shirt and khaki pants, running shoes
- suspect had a noticeable scar on his left cheek

Scenario Two: Victim B

You are Darlene Lewis, living in Apartment 102 at 456 Bloor Street. Your neighbour, Marla Kindra, in Apartment 104, has been causing you problems. Marla often goes on drinking binges in the late hours and then begins to harass you. Tonight, Saturday, February 18, 2006, at 0130, she comes knocking at your door, screaming to be let in. When you open the door, she is noticeably intoxicated and wants to argue with you. She calls you names and accuses you of stealing her groceries. You deny her accusations. You have not stolen her groceries, and you have never even been in her apartment before. When you ask Marla to leave, she refuses. At this point, you believe she can cause you harm, so you leave your apartment and go to a neighbour's place to phone the police. While you are phoning, the neighbour, Jack Toselles, goes to check on Marla and discovers that she has gone back to her apartment. When the police arrive on the scene, you tell them what happened that night.

Scenario Three: Witness A

You are Lydia Agnew, working as a clerk at Jewels Etc. in the Anywhere Mall. It is 1830 on March 14, 2006. The store has been fairly busy, and you have been spending most of your shift at the cash register. As you are assisting a customer with a purchase, you glance over at the sunglasses display and notice a teenage girl slip a pair of expensive sunglasses into her purse. You immediately run over and ask the girl to come with you to the back of the store. She refuses and says she has not done anything wrong. You and another employee escort her to the back room and phone the police. The sunglasses are found in her purse, and she is charged with possession of stolen property under $500.

> Suspect's name — Shelley Bertrand   Parent—Florence Bertrand
>
> Age — 16
>
> Birthdate — May 15, 1989
>
> Took a pair of sunglasses valued at $157.95

Scenario Four: Witness B

You are Ernest Spadaforia, and you have recently moved into a new neighbourhood. You have been noticing some strange things happening at the house directly across from yours. You have suspicions that the residents in the house are selling drugs. You often notice people drive up to the house, give the residents money, and receive small bags of something. You have never seen anything close up, but it always seems that the same thing is being exchanged. On a few occasions, you have also noticed the residents smoking a joint while sitting on their front steps.

ACTIVITY: WRITING POLICE NARRATIVES

The following scenarios are scrambled. Write the police narrative by placing the reports in chronological order. Also eliminate any subjective or irrelevant information. The information that belongs in the introduction and conclusion of the police narrative is not included. Be sure to include that information when you are writing the final report.

Scenario One

The station wagon was seen leaving the parking lot at 0045. The suspect fled in an unknown direction. The suspect held the cashier at knifepoint. Constable Sewer and

Constable Coney contained the area and the K-9 unit was called. Anne Stevens appeared to be upset about the incident. Anne Stevens saw a light-brown station wagon leaving the area, but it is not known if it is connected to the robbery. The suspect took $500.00 from the cash register. Around 0030, a white male walked into Pluff's Pizzeria at 514 Anywhere Road with a stocking over his head. The suspect was also holding a knife. The K-9 unit and the officers had no success in capturing the suspect. Black Sabbath music was playing when the officers arrived at the scene. The constables questioned Anne Stevens, who lives above the pizzeria. Joey Wong was working at Pluff's Pizzeria when the suspect approached him.

Scenario Two

The incident happened at approximately 1500. The Extunes own a beautiful 3000-square-foot home in the country. Mr. Extunes stated that he was afraid when he saw the accused trying to gain entry into his home. The accused attempted to break into the home of Mariah and Phillipe Extunes of 1100 Anywhere Street on October 11, 2006. When Mr. Extunes went to his kitchen, he noticed the accused using a crowbar and attempting to pry the patio door open. Mr. Extunes called the police, and the K-9 unit was able to track Salem Beijing hiding in the nearby woods. The accused has blond streaks throughout his black hair. Salem Beijing was arrested for attempted break and enter and possession of housebreaking tools. The accused was wearing a black turtleneck and black jeans when the incident occurred. Mr. Extunes stated that he heard noise coming from the patio door of the kitchen, and he went to see what was causing the noise. Mrs. Extunes was not home when the incident happened. Some damage appears to have been done to the patio door. Mr. Extunes approached the patio door, and when the accused saw Mr. Extunes, he fled. Mr. Extunes stated that he had never seen the accused before.

Scenario Three

When Constable Balsh approached Mr. Blake, Mr. Blake appeared to be intoxicated. Mr. Blake was charged with assaulting police and carrying a concealed weapon. Upon arriving at the scene, Maxine Lethbridge greeted Constables Balsh and Simong at the door and stated that a drunken patron would not leave the premises. Maxine Lethbridge, the bartender, was wearing a red leather outfit. Ms. Lethbridge told the constables that after serving the patron eight drinks, she told the patron she would not serve him any more alcoholic beverages. Mr. Blake was drinking Singapore Slings. Constable Simong restrained Mr. Blake. When Constable Balsh approached Mr. Blake, Mr. Blake was slurring his words, and he was unsteady on his feet. Ms. Lethbridge asked Mr. Blake to leave the premises and Mr. Blake refused to leave. Ms. Lethbridge noticed that after several drinks, Mr. Blake was slurring his words and that he staggered as he walked to the restroom. When Constable Balsh asked Mr. Blake to leave the premises, Mr. Blake kicked Constable Balsh in the left leg several times and attempted to punch Constable Balsh. Bill Blake became angry with Ms. Lethbridge and continued to ask for more drinks. Ms. Lethbridge stated that Mr. Blake began swearing at Ms. Lethbridge when she refused to serve him, and she then called the police for assistance. Mr. Blake was arrested and a search netted a concealed knife in the back of his pants. Mr. Blake was wearing a pair of khaki cargo pants. Mr. Blake was charged with assaulting police and carrying a concealed weapon.

Scenario Four

Mr. Chou asked Mr. Peltier to leave for a second time. When Mr. Chou and his friends left the bar to go back to Mr. Chou's place, Mr. Peltier left with them. Upon arrival at the scene, the complainant, Louie Chou, greeted Constables Migwan and Brisque. Mr. Chou told the officers that Mr. Richard Peltier assaulted him by hitting him over the head with a beer bottle. When the group of friends arrived at Mr. Chou's house, Mr. Peltier continued to drink beer. Mr. Peltier became annoyed with Mr. Chou's request, because he wanted to stay and continue partying. Mr. Chou told the officers that earlier that evening he and his friends went to Chelsey's Roadhouse for a few drinks, and they met an acquaintance, Richard Peltier, there. At 0030, Mr. Chou asked his visitors to leave because he had to work in the morning. Mr. Chou stated that Mr. Peltier was intoxicated at Chelsey's when they met him there. Once Mr. Chou asked his guests to leave for a second time, Mr. Peltier took the beer bottle and struck him over the head until the bottle broke. There was glass everywhere on the floor. Mr. Chou's head was bleeding. At this point, Mr. Chou called the police. After Mr. Peltier struck Mr. Chou with the beer bottle, he fled from the residence. Constable Migwan and Constable Brisque escorted Mr. Chou to Anywhere Hospital where he obtained five stitches for the cut he received on the head.

Scenario Five

After interviewing Mr. Sombers and Ms. Brownings, Constables Eshanagan and Romano discovered that Mr. Sombers was driving westbound on Highway 17, crossed over the centre line, and struck Ms. Brownings, who was driving eastbound on Highway 17. Mr. Sombers was on his way to Somewhere City. The accident occurred on Highway 17, near Obasango Park, which is 40 kilometres from Anywhere City. Both Mr. Sombers and Ms. Brownings indicated that no injuries occurred. Upon arriving at the scene, Constables Eshanagan and Romano noticed the front end of a black Honda Civic dented in and the front end of a 1998 silver Pontiac Sunfire dented. Ms. Brownings was upset that her car was hit. Constable Romano called a tow truck to take both vehicles to Jack's Garage on 145 Anywhere Street. Constables Eshanagan and Romano were dispatched to a report of a two-vehicle highway accident. Passengers were waiting outside the vehicles when the officers arrived at the scene, and they appeared uninjured. The driver of the Pontiac Sunfire was Melvin Sombers of 40 Anywhere Street, Anywhere, Ontario, and the driver of the Honda Civic was Melissa Brownings of 456 Somewhere Street, Somewhere, Ontario. The Pontiac Sunfire was a nicer car. Mr. Sombers was charged with careless driving.

SPECIFIC POLICE REPORTS

General Occurrence Reports

The general occurrence report is the standard police report that is written for many different complaints such as a break and enter, an assault, or a car theft. It is written when a complainant or a victim reports an incident to the police.

There are two parts to the general occurrence report. The first part of the form requires the officer to fill in the blank spaces. The officer will fill in information such as the incident type, time, date, and location of the incident, name of the officer, and the report date and time. The second part of the form is the narrative. This is the more difficult of the two to complete. This is where the officer describes in full the facts and circumstances that make up the incident.

Constable Mashetta was dispatched to a break-and-enter report. Here is an example of her completed General Occurrence Report:

Metropolitan Police Department

General Occurrence Report

Officer Badge No.: 0125   Officer Name: Constable Mariah Mashetta

Incident No.: 234

Incident Type: Break and Enter

Incident Date: October 29, 2005   Time: 0125

Report Date: October 29, 2005   Time: 0330

Details How Committed: It appears that the suspect used an axe to gain entry through the front door

Weapons or Tools Used: It appears that an axe was used

Property General Description: Sony DVD player - $500.00, 14-karat diamond ring - $2500.00

Value of Loss: $3000.00   Value of Damage: Door - $2000.00

Recovery Value: $3500.00

REPORT NARRATIVE

On Tuesday, October 29, 2005, at 0115, Constable Maschetta was dispatched to 120 Anywhere Street to take a report on a break and enter.

The complainant Mrs. Shaylea Mosiak, owner of the residence, met Constable Maschetta. Mrs. Mosiak told Constable Maschetta that she left her residence at 2130 that evening to attend a dinner dance. She was out the entire evening and returned at approximately 0100. As she drove into her driveway, she noticed that the front door of her house was opened. As she approached her house she noticed the door was smashed open. Mrs. Mosiak stated that after examining the house, she discovered her Sony DVD player valued at $500.00 was taken from her living room and her 14-karat diamond ring valued at $2500.00 was missing from her bedroom jewellery box. Everything else appeared to be untouched.

Constable Maschetta examined the front door, and she believes an axe was used to gain entry into the house.

Constable Maschetta advised Mrs. Mosiak to contact her insurance company and to call the station if she discovered anything else missing. Constable Maschetta left the scene at 0210.

PRACTICE: WRITING GENERAL OCCURRENCE REPORTS

Scenario One

You are a constable working at the Metropolitan Police Department. You are dispatched to a break and enter at 41 Anywhere Drive at 0130 on March 14, 2006. The complainant, Louigi Lasterommi, tells you that his 26-inch Sony television, Panasonic DVD player, Sony PlayStation, and IBM computer have been stolen. The value of the merchandise is estimated to be $2600.00. It appears that the suspect gained entry through the kitchen window that was left open. Complete the general occurrence report for this incident.

Scenario Two

You are a constable working at the Metropolitan Police Department. It is May 15, 2006, and you are dispatched to take a report on a noise complaint at 770 Queen Street at 0230. The complainant, Quinton Masters, tells you the loud music and partying occurring next door is keeping him up. You talk to the neighbour, Leroy Brown, and he turns the music off and sends his party guests home. Write the general occurrence report for this incident.

Scenario Three

You are a constable working at the Metropolitan Police Department. It is September 10, 2006, and you are dispatched to take a report on a possible stalking. You go to 132 Sussex Drive at 1630 and you talk to Fred Flint. He tells you that a neighbour, Salina Strustrum, is following his son, Billy Flint. Billy is 12 years old and Salina Strustrum is 41 years old. Fred has talked to Salina and has asked her to stop following Billy, but she has not. Billy is beginning to get nervous about what is happening. Today, on his way home from school, Billy noticed Salina following behind him and trying to hide behind trees. Write the general occurrence report for this incident.

Supplemental Report

The supplemental report is used if additional information is discovered about an incident that was already reported. For example, if a break and enter occurred, the officer would complete the general occurrence report. If, later on, subsequent information about the incident is discovered, a supplemental report is written. The same report number is given on the supplemental report as was originally given on the general occurrence report. The officer taking the report should include the complainant's name, address, and nature of the complaint on the report. To avoid any possible confusion, the officer may begin the narrative by stating, "Further to the break-and-enter report, dated October 29, 2006, that occurred at 120 Anywhere Street ... "

Additional information regarding the break and enter that occurred at 120 Anywhere Street was discovered. Here is the supplemental report that was written to document the subsequent information:

Metropolitan Police Department

Supplemental Report

Officer Badge No.: 425   Officer Name: Constable Valerie Valley

Incident No.: 234

Incident Date: October 31, 2006   Time: 1530

Report Date: October 31, 2006   Time: 1700

REPORT NARRATIVE

Further to the report of the break and enter at 120 Anywhere Street, dated October 29, 2006, Constable Valley was dispatched to Toby's Pawn Shop, 289 Somewhere Street, at 1530 to take a report on information regarding the break and enter.

Toby Tallahackie, owner of Toby's Pawn Shop, phone # 235-6915, greeted Constable Valley. Mr. Tallahackie told Constable Valley that he believed he had some information regarding the stolen merchandise from the break and enter that had occurred a few days ago. Mr. Tallahackie called the police after he had watched the news and heard that a 2-karat diamond ring was taken from the residence. Mr. Tallahackie stated that a man visited
his pawn shop on October 30, 2006, and sold him a 2-karat diamond ring for $850.00.

Mr. Tallahackie provided Constable Valley with the name and address of the individual who sold him the diamond ring. The individual was Blasher Brashid of 91 Trouble Street, Anywhere, Ontario.

Constable Valley left the premises at 1610. Further investigation will continue.

PRACTICE: WRITING SUPPLEMENTAL REPORTS

Scenario One

You are a constable working at the Metropolitan Police Department. It is March 14, 2006, and you are dispatched to 41 Anywhere Drive at 1000 to take a report on information regarding the break and enter that had occurred at the residence. Louigi Lasterommi tells you that he has also discovered that some of his tools are missing. His Black and Decker skill saw and drill that were in his shed are gone. It appears that the padlock on the shed door has been cut off. The items are valued at approximately $250.00.

Scenario Two

You are a constable working at the Metropolitan Police Department. It is May 15, 2006, at 0330 and you are dispatched to 770 Queen Street to take a report on a noise complaint. Quinton Masters had called in earlier to report a noise complaint. When you arrive at his residence, he tells you that the neighbour, Leroy Brown, has started the music and partying once again. You go to talk to Mr. Brown once again, and he assures you that he will stop the music and partying and will not cause any other problems. Write the supplemental report for this incident.

Scenario Three

You are a constable working at the Metropolitan Police Department. It is September 12, 2006, at 1700, and you are dispatched to 132 Sussex to take additional information regarding a stalking. Fred Flint greets you. He tells you that the neighbour, Salina Strustrum, is still bothering his son, Billy. Even though the police had talked to her, she followed Billy home from school today. Her behaviour is frightening Billy. Write the supplemental report for this incident.

Arrest Report

The arrest report consists of two parts as well. The top portion of the report requires that the arresting officer fill in the blanks. The bottom portion of the report requires that the officer describe the incident and justify the arrest. Sufficient evidence justifying the arrest must appear in the narrative. If the arrest occurs following an incident previously documented, the incident number on the arrest report should remain the same. The officer may want to begin the report by noting "Further to the report of the break and enter on 120 Anywhere Street, dated October 29, 2006 . . . "

Here is an example of a completed arrest report:

Metropolitan Police Department

Arrest Report

Officer Badge No.: 425   Officer Name: Constable Valerie Valley

Incident No.: 1245

Surname of Arrested Person: Brashid   Date of Birth: September 15, 1960

Given Name 1: Blasher   Given Name 2: Viallin   Sex: M

Incident Date: October 29, 2006   Time: 0125

Report Date: October 31, 2006   Time: 2000

Arrest Date: October 31, 2006   Time: 1715

Place of Arrest: 91 Trouble Street   By Officer #: Constable Valley #425

ARREST NARRATIVE

Further to the report of the break and enter dated October 29, 2006, incident number 1245, Constable Valley obtained a search warrant and proceeded to 91 Trouble Street, residence of Blasher Brashid, at 1630.

Upon arrival at Mr. Brashid's residence, Mr. Brashid met Constable Valley at the front door. Constable Valley questioned Mr. Brashid regarding the diamond ring he sold to Toby's Pawn Shop. Mr. Brashid told Constable Valley that he purchased the ring at Diamonds Forever for a former fiancee. Mr. Brashid said he had no use for the ring any longer, so he decided to sell it to the pawn shop. Mr. Brashid stated that he purchased the ring six months ago, and he no longer had the receipt. When Constable Valley questioned Mr. Brashid as to his whereabouts on the night of October 29, 2006, Mr. Brashid stated that he was home by himself.

Constable Valley presented Mr. Brashid with the search warrant and proceeded to search the house for the Sony DVD player. Nothing was found. Constable Valley then searched the garage and discovered a black tarp in the back left-side corner of the garage. Constable Valley discovered that under the tarp were many items including a Sony DVD player. Other items found included a 25-inch JVC television, a Sony video camera, and an Alpine car stereo.

Constable Valley read Mr. Brashid his rights and arrested him for possession of stolen property.

Constable Valley left the scene at 1715 and escorted Mr. Brashid to the Metropolitan Police Department for containment.

PRACTICE: WRITING ARREST REPORTS

Scenario One

You are a constable working at the Metropolitan Police Department. It is March 25, 2006, at 0030. You are patrolling your assigned area when you notice a car on Careful Avenue swerving in and out of lanes. You stop the vehicle, and the driver appears to be intoxicated. You can smell alcohol on her breath. After administering a Breathalyzer test, you find the results are well over the legal limit. You arrest Victoria Valente for driving while under the influence of alcohol. Write the arrest report for this incident.

Scenario Two

You are a constable working at the Metropolitan Police Department. It is June 30, 2006, at 2200, and you are dispatched to Bullrider's Bar at 105 Wilson Street to take a report on a bar fight. You arrive at the bar and discover the altercation in progress. Sam Statson is standing over Vladamir Szeuopod, and he is punching him continually in the face. Vladamir Szeuopod is bleeding and he appears to be unconscious. Once you restrain Sam Statson and call an ambulance, you arrest Statson for assault causing bodily harm. Write the arrest report for this incident.

Scenario Three

You are a constable working at the Metropolitan Police Department. It is July 7, 2006, at 2130, and you are dispatched to a domestic dispute at 415 Megginson Avenue. When you arrive at the residence, you discover that Mrs. Nancy Lethbridge, complainant, has a black eye and a bleeding lip. She tells you that her son, Jonah Lethbridge, physically assaulted her. Mrs. Lethbridge stated that she confronted her son about his drug problem, and he began hitting her. Her son has locked himself in his bedroom. You arrest him for assault causing bodily harm. Write the arrest report for this incident.

DISCUSSION QUESTIONS

1. List three types of reports that a police officer might write.
2. List three organizations other than the police services that might read police reports.
3. List two benefits that police reports have to outside agencies or persons.
4. List the characteristics of a well-written report.

ACTIVITY: WRITING GENERAL OCCURRENCE REPORTS, SUPPLEMENTAL REPORTS, AND ARREST REPORTS

General Occurrence Reports

Scenario One

You are a constable working at the Metropolitan Police Department. It is February 5, 2006, at 1630, and you are dispatched to a break and enter at 10 Somewhere Street. Melvin Peltune, who tells you that his shed was broken into and that his snow blower was stolen, greets you. The shed had a padlock on it. After examining the padlock, you believe it was cut open. The snow blower is a Honda, 200 horsepower, valued at $2150.00. Write the general occurrence report for this incident.

Scenario Two

You are a constable working at the Metropolitan Police Department. It is August 20, 2006, at 0800, and you are dispatched to take a report on vandalism at 605 Loretta Street. Brian Van Gough greets you, and he tells you that someone has vandalized his car. The car, a white 1998 Cutlass Oldsmobile, has been spray-painted black in various places. The windows, door panels, hood, and trunk are all covered with black paint. The tires have been slashed. This incident occurred sometime during the night. Write the general occurrence report for this incident.

Scenario Three

You are a constable working at the Metropolitan Police Department. It is February 5, 2006, at 1930, and you are dispatched to a car accident on Queen Street East. A snowstorm is in progress, and driving conditions are poor. You arrive at the scene, and you notice a blue 2000 Ford half-ton truck and a 1996 grey Dodge Caravan at the lights. The Caravan's back bumper is smashed in and the Ford truck appears to have scratches on the front bumper. The Caravan is sitting directly in front of the truck. It appears that the truck hit the Caravan from behind. Skid marks are apparent in the snow. You talk to Cee Zwang and Caroline Hight. They tell you that Cee Zwang, owner of the Caravan, was sitting at the red light when she was hit from behind. Caroline Hight, owner of the truck, stated that she tried to stop when she saw the red light, but her truck slid, and she couldn't stop. Write the general occurrence report for this accident.

Scenario Four

You are a constable working at the Metropolitan Police Department. You are dispatched to a domestic dispute at 330 Northern Street. The victim, Simone Levesque, tells you that her boyfriend, René Pitoux, assaulted her. He grabbed her by the throat and struck her on the head a number of times during an argument. The victim had bruises on the neck and some bleeding on the scalp. You call the Women's Crisis Centre for assistance.

Scenario Five

You are a constable working at the Metropolitan Police Department. You are dispatched to the Metropolitan Mall, 123 Main Street, to take a report on a stolen car. You meet Lila McCormick in the parking lot, and she tells you the following: She parked her vehicle in the lot at 0930 hours in the morning. The car doors were locked when she exited the car. When she finished her shopping and returned to her car, it was not there. Thinking she had forgotten where she parked it, Lila looked around, but it was nowhere to be found. Ms. McCormick stated that her car is a 2003 Pontiac Sunfire. The car is silver in colour. Her licence plate number is 800 YME. Ms. McCormick has insurance on the vehicle.

Supplemental Report

Scenario One

You are a constable working at the Metropolitan Police Department. It is February 5, 2006, at 1800, and you are dispatched to 10 Somewhere Street. You had been there earlier to take a report on a break and enter. The complainant, Melvin Peltune, had reported that his snow blower was stolen from the shed. When you arrive at Mr. Peltune's residence, he tells you that after further investigation, he noticed that his basement window was smashed. He believes the suspect had also tried to gain entry to his basement and was unsuccessful. You take a look at the window and discover the window is smashed in various places. It appears that someone used something hard, such as a rock or bat, to try to break the window. Write the supplemental report for this incident.

Scenario Two

You are a constable working at the Metropolitan Police Department. It is November 22, 2006, at 1500, and you are dispatched to take a report on possible information about a car theft. Last week, a number of Ford Windstar vans were taken from homes around the city. When you arrive at 21 Paladin, you are greeted by Jake Williams. He tells you that his friend, Rudy Writhe, is responsible for the thefts. He states that his friend is planning to repaint the vans and sell them to someone in another city who sells stolen property. Write the supplemental report for this incident.

Scenario Three

You are a constable working at the Metropolitan Police Department. It is February 5, 2006, at 2100, and you are dispatched to 115 Somewhere Street to take information regarding a car accident involving a 1996 Dodge Caravan and a 2000 Ford half ton. Cee Zwang, owner of the Caravan, was hit from behind. She has called you because she now claims that her back and neck are extremely sore. She wants it documented that she is now in pain due to the accident. You tell Cee Zwang to go to the hospital. Write the supplemental report for this incident.

Scenario Four

You are a constable working at the Metropolitan Police Department. You are dispatched to take a report on the possible location of a stolen vehicle. Last week a report was taken on a stolen silver Pontiac Sunfire from Metropolitan Mall. You arrive at Metropolitan High School and are greeted by Matthew Mastriolli, principal of the school. You notice in the parking lot there is a Pontiac Sunfire that fits the description of the car that was stolen the previous week. Mr. Mastriolli tells you that he has noticed that the vehicle has been parked in the school lot and has not been moved. The vehicle has been parked there for four days. Mr. Mastriolli said the car does not appear to belong to anyone at the school. The licence plate number of the vehicle is 800 YME.

Arrest Report

Scenario One

You are a constable working at the Metropolitan Police Department. It is December 13, 2005, at 2000, and you are dispatched to Continuous Accessories, a store in the Metropolitan Mall, to take a report on a theft. You arrive at the store and the store manager, Sarah Lu, has restrained Bethany Miller, age 16, in the store. Bethany was shoplifting and was attempting to take a leather wallet, valued at $60.00. Sarah saw Bethany put the wallet inside her jacket. You charge Bethany with theft under $500.00.

Scenario Two

You are a constable working at the Metropolitan Police Department. It is April 7, 2005, at 2230, and you are dispatched to Reggie's Salon regarding a drunken patron. You arrive at the bar and are greeted by Sal Ocha, bartender, who tells you that a patron needs to be escorted from the bar. The patron, Bob Boyd, has had too much to drink, and refuses to leave. When you approach Mr. Boyd, he starts to kick and throw punches at you. After restraining him, you charge him with assaulting an officer. Write the arrest report for the incident.

Scenario Three

You are a constable working at the Metropolitan Police Department. It is July 12, 2005, at 2000, and you are patrolling your given area. You notice a 1997 Honda Accord speed by on Main Street. You chase the car and manage to get the driver to pull over. When you talk to the driver, you notice an odour coming from the car that smells like cannabis. On searching the vehicle, you find 15 grams of cannabis. You arrest Joe Smith, owner of the vehicle, for possession of narcotics. Write the arrest report for the incident.

Scenario Four

You are a constable working at the Metropolitan Police Department. It is September 15, 2005, and you have just come from taking a report on an assault. Simone Levesque had reported that her boyfriend, René Pitoux, assaulted her. She had bruises on her neck and bleeding on her scalp. Simone and her boyfriend had an argument when the incident happened. Simone had told you that this was not the first time that her boyfriend had assaulted her. Ms. Levesque gave you Mr. Pitoux's address. You go the Mr. Pitoux's residence at 476 Bloor Avenue and arrest him for assault causing bodily harm.

WRITTEN COMMUNICATIONS COMPETENCY EXAMS

As you finish your courses of study in law enforcement, you will begin to think about applying for positions as a police officer. Many police departments will require that you take a written communications competency exam. The following scenarios are provided to help you practise and prepare for the written communications competency exam.

The goal is to prepare a written report/narrative for your supervisor. The information provided in each scenario is not organized. You must pull out all the facts and organize them. When you are completing these practice scenarios, remember the following:

✓ Be aware of the time you will be given to write the exam. If you are given 60 minutes, use your time effectively to ensure you finish within this time.

✓ Read the scenario very carefully and list in point form all the facts that you think are important. Use the headings *Time*, *Location*, and *Evidence*. Eliminate any subjective unimportant information from the scenario.

✓ From your list of facts, write a narrative summarizing what happened. Remember to use the facts that you have listed.

✓ Draw a conclusion or interpretation as to what happened in this incident.

✓ Check your spelling and grammar.

Scenario One

Sally Sassoon was travelling west on Highway 207. Billy Billoux was driving east and attempted a right turn southbound into Highway 459 from 207. Ms. Sassoon's car slid into the southbound lane. The roads were slippery as a result of the snowstorm the previous evening. The accident occurred at the intersection of Highway 207 and Highway 459, about 5 km west of Somewhere, Ontario. Mr. Billoux's car skidded into a Ford half-ton truck owned by Josie Jiles, which was facing north on Highway 459, waiting at the intersection to enter Highway 207. Ms. Sassoon was approaching the intersection at Highway 459 and was stopped and waiting for clear roads to make the turn. Mr. Billoux was listening to the radio while he was driving. Ms. Sassoon was taken to St. Mary's Hospital with a broken knee and a possible concussion. Ms. Sassoon is upset about the accident. Ms. Jiles is related to Chief Edwards. Ms. Sassoon's car has extensive damage and was towed to Art's Auto Body, 1330 Fixit Street. Mr. Billoux's car has damage to the front and back end. Josie's vehicle had damage to the front end. The accident occurred at 0830 hours. All parties involved in the accident were on their way to work. The impact of the accident caused Billy's car to rear end and swing into the westbound lane of 207 where Sally's car was stopped.

List all of the facts, under the following headings: *Time*, *Location*, and *Evidence*. Using the facts listed above, write the narrative for the report. Remember to draw a conclusion or interpretation.

Scenario Two

Edward Smith was on his way to work at ABC Construction. Two other vehicles stopped to see if everyone was okay. Luc Jones's car is a 2000 Ford Taurus. The accident occurred at 0700 hours. Mr. Jones was approaching the corner of Queen Street and Pine Street and put on his brakes to stop for a red light. Skid marks made from Mr. Smith's car measured three metres. Mr. Jones was on his way to work at Alley's Computers. Mr. Smith's car is a 1999 Grand Am Pontiac. Mr. Jones reported that he was listening to the news on the radio when the accident occurred. Mr. Jones's car slid when he braked, and it went through the red light. Mr. Smith was listening to an Aerosmith CD when the accident occurred. Mr. Smith tried to swerve to miss Mr. Jones's car, but he was not successful. The skid marks from Mr. Jones's car measured 1.5 metres. Mr. Jones was travelling east on Queen Street. Mr. Smith's car was travelling north on Pine Street and hit Mr. Jones's car. The accident occurred three miles from Mr. Smith's house. The roads were slippery due to a snowstorm the previous evening. Mr. Jones used his cell phone to call the police to report the accident. Mr. Smith indicated that he owned a cell phone.

List all of the facts, under the following headings: *Time*, *Location*, and *Evidence*. Using the facts listed above, write the narrative for the report. Remember to draw a conclusion or interpretation.

Sample Templates: For more blank templates to practise on, please visit www.pearsoned.ca/turpin.

Pages 126 through 128 show actual examples of the general occurrence report, the supplemental report, and the arrest report from the Sault Ste. Marie Police Service. The Sault Ste. Marie Service is one of 41 services within Ontario that are part of the Ontario Police Technology Information Centre (OPTIC).

To see what other departments are part of this system and to see other examples of the templates used by OPTIC, please visit www.pearsoned.ca/turpin.

See page 126 for an example of what an actual general occurrence report looks like from the Sault Ste. Marie Police Service.

An example of what an actual supplemental report looks like, also from the Sault Ste. Marie Police Service appears on page 127.

For an example of what an actual arrest report looks like, again from the Sault Ste. Marie Police Service, see page 128. This represents only one page of the report. Please refer to www.pearsoned.ca/text/turpin for the full 4-page report.

SAULT STE. MARIE POLICE SERVICE
DATA ENTRY GUIDE

GENERAL REPORT:

OCCURRENCE #:	**TASK #:**
AUTHOR: (DICTATING OFFICER)	**REPORT TIME: (TIME OF DICTATION)**
ENTERED BY: (TRANSCRIBER)	**ENTERED TIME: (DEFAULTS TO PRESENT TIME)**
REMARKS:	

NARRATIVE:

(TRANSCRIBE GENERAL REPORT NARRATIVE AS DICTATED)

SAULT STE. MARIE POLICE SERVICE
DATA ENTRY GUIDE

SUPPLEMENTARY REPORT:

OCCURRENCE #:	**TASK #:**
AUTHOR: (DICTATING OFFICER)	**REPORT TIME: (TIME OF DICTATION)**
ENTERED BY: (TRANSCRIBER)	**ENTERED TIME: (DEFAULTS TO PRESENT TIME)**
REMARKS:	

NARRATIVE:

(TRANSCRIBE SUPPLEMENTARY REPORT NARRATIVE AS DICTATED)

SAULT STE MARIE POLICE SERVICE
ARREST REPORT

S/SGT.	☐
DUTY OFFICER	☐
CERB	☐
SHIFT CLERK	☐

☐ ADULT	☐ MALE ☐ FEMALE	CELL NO. ☐ ☐

ARRESTING OFFICER:	ARRESTING OFFICER:	INVESTIGATING OFFICER:
BADGE #:	BADGE #:	BADGE #:

OCCURRENCE ADDRESS:	DOMESTIC VIOLENCE INCIDENT ☐

CHARGES:

OCCURRENCE #:	CHARGE:	OCC DATE:
OCCURRENCE #:	CHARGE:	OCC DATE:
OCCURRENCE #:	CHARGE:	OCC DATE:
OCCURRENCE #:	CHARGE:	OCC DATE:
OCCURRENCE #:	CHARGE:	OCC DATE:

NAME TYPE:

☐ PRIMARY ☐ ALIAS	☐ NICKNAME ☐ MAIDEN NAME	☐ VARIANT ☐ OTHER	☐ ACRONYM ☐ LEGAL	☐	OPERATING

ACCUSED PERSON:

SURNAME:	G1:	G2:	G3:
SEX:	DOB:	(OR AGE):	

NAME TYPE:

☐ PRIMARY ☐ ALIAS	☐ NICKNAME ☐ MAIDEN NAME	☐ VARIANT ☐ OTHER	☐ ACRONYM ☐ LEGAL	☐	OPERATING

ALIAS:	NICKNAME:

ADDRESS TYPE:

☐ RESIDENCE ☐ SECOND/SEASONAL RESIDENCE ☐ TEMPORARY RESIDENCE	☐ BUSINESS ☐ NO FIXED ADDRESS ☐ FREQUENTS	☐ OBSERVED ☐ CHECKED ☐ OTHER (CLARIFY)

ST/FIRE #:	STREET:	TYPE:	DIRECTION:
CITY:		PROVINCE:	

☐ APT. ☐ SUITE ☐ UNIT #	COMMON NAME:

LOT:	CONC:	SITE:
P.O. BOX #	ROUTE:	☐ R.R. ☐ S.S. ☐ #

ADDRESS HAZARD:

☐ ANIMAL ☐ FAMILY VIOLENCE	☐ CLUBHOUSE ☐ OCCUPANT	☐ DANGEROUS GOODS ☐ WEAPON	☐ EXPLOSIVES ☐ OTHER (CLARIFY)

TELEPHONE NUMBER:

1. ( )	TYPE:	2. ( )	TYPE:

E-MAIL ADDRESS:

CONDITION OF ACCUSED:

SOBER ☐	INTOX ☐	HAD BEEN DRINKING ☐	UNDER INFLUENCE DRUGS ☐	MENTAL INSTABILITY ☐	EMOTIONAL ☐

CAUTION: | REMARKS:

☐ ARMED & DANGEROUS ☐ CONTAGIOUS DISEASE ☐ HATES POLICE ☐ NON-CONTACT/COMM ☐ STALKER ☐ WEAPON USED	☐ ALCOHOL ☐ DRUGS ☐ HIGH RISK OFFENDER ☐ RESISTS ARREST ☐ SUICIDE RISK ☐ OTHER (CLARIFY)	☐ ATTEMPT SUICIDE ☐ ESCAPE RISK ☐ MEDICAL ☐ SEX OFFENDER ☐ UNPREDICTABLE BEHAVIOR	☐ CARRIES WEAPONS ☐ FAMILY VIOLENCE ☐ MENTALLY DISORDERED ☐ SOLVENT ABUSE ☐ VIOLENT OR ASSAULTIVE

FPS #:	D/L #:	PROVINCE:
BIRTHPLACE:	PROV:	COUNTRY:	CITIZENSHIP:
EMPLOYER:		OCCUPATION:	

Revised April 2004

Public Speaking

LEARNING OUTCOMES

Upon successful completion of this chapter, you should be able to

1. describe and use the different stages involved in preparing your speech
2. describe and use the different aspects involved in delivering your speech
3. recognize the importance of knowing your audience
4. use visual aids effectively in your speech
5. deliver an effective speech
6. describe and use procedures you should follow after the speech is finished
7. effectively evaluate your speech

GIVING A SPEECH

As a police officer you may be required, at times, to give a speech as part of your job. For example, your staff sergeant may ask you to give a speech to a group of people in a community regarding the neighbourhood watch program, or you may be called upon by a college to speak to a group of students about the law enforcement profession. You may also be part of a special team at work that may involve giving a speech to your fellow officers or superiors regarding your project. Giving speeches is not everyone's favourite thing to do. Some people have not had much practice giving speeches and may be nervous about doing so. Just keep in mind that becoming a good public speaker takes practice, like everything else. Following a certain procedure will help you become a confident speaker.

STAGES

When you are giving a speech, there are four stages you should follow. Following through with each of these stages will ensure that you deliver your best product. The stages are as follows:

1. Organizing the content for your speech
2. Preparing for your speech
3. Delivering your speech
4. Evaluating your speech

1. ORGANIZING YOUR CONTENT

Your speech will contain an **introduction**, **body**, and **conclusion**.

In your introduction, you want do two things. The first is to start with several sentences that will engage your audience's attention. You want your audience to get hooked into what you are going to talk about. You can engage their attention by telling a brief story that relates to your topic, or you can start off with several related questions. The key is to start off in a way that will keep the audience's attention. The second thing you want to do in your introduction is make your main point. This is your thesis statement. Tell the reader in your thesis statement what your subject is and what point you want to make about your subject.

In the body of your speech, you provide the facts, evidence, and support to help the audience see exactly what you mean by your thesis statement. The body provides the meat of the presentation. After you have finished presenting the body, the audience will have a clear picture of what you wanted to say.

In the conclusion, you do more than just put a finish on your speech. You do two things. First, you restate your thesis. Tell the audience again what your main point is. Second, present an ending that will keep your audience thinking about your subject. You do not want your audience to leave without reflecting on what you had to say. You may accomplish this by posing a few questions related to your subject before closing or by making some recommendations.

SKETCH OF THE CONTENT OF A SPEECH

> **Introduction**
> Attract your audience's attention.
> Give a thesis statement.

> **Body**
> Give the facts, evidence, and support for your thesis statement.

> **Conclusion**
> Reiterate your thesis statement.
> End in a way that will keep your audience thinking about your subject.

PRACTICE: WRITING AN INTRODUCTION

Assume you are a constable working for the Metropolitan Police Department. You have been asked by the coordinator of the law enforcement program at your community college to speak to the students about a career in law enforcement. You are working on your introduction. Prepare a thesis statement for this subject. Then write the opening remarks for the thesis statement. You want these opening remarks to be interesting and to capture the attention of your audience.

USING AN OUTLINE FOR THE PRESENTATION

When you are preparing your speech, you should use an outline to help you organize your thoughts and information. Also, using an outline will provide you with the correct structure for your speech. When you give your speech, your audience will be able to follow what you are saying. When you are filling in your outline, remember that it is only a guide. Use only key words that will guide you when you are actually writing.

PRACTICE: USING AN OUTLINE

Assume you are a constable working for the Metropolitan Police Department, and you are preparing to give a speech to law enforcement students about a career in policing. Using an outline, jot down the points you would make to the students.

WRITING THE CONTENT OF YOUR SPEECH

Once you have completed an outline, you are ready to compose the content of your speech.

SAMPLE OUTLINE

Introduction

Opening remarks _____
_____
_____

Thesis statement _____
_____
_____

Body

Paragraph one support _____
_____
_____
_____
_____

Paragraph two support _____
_____
_____
_____
_____

Paragraph three support _____
_____
_____
_____
_____

Conclusion

Restate thesis statement _____
_____

Closing statements _____
_____
_____

At this point, you may want to write out the contents in full so that you know exactly where you are heading. Remember that with any type of writing, your first draft will not be your last; this applies to writing the details of the speech. Your first draft is just that. You will make changes to it and rewrite it until you are satisfied with a final copy.

When you are finished drafting the details in full, create note cards that you can use when you are presenting. You will not want to take your written text and read directly from it. Create note cards with information that you can use to guide you along with your speech and keep you on track. During your presentation, you can refer to your note cards to lead you to your next point. It is critical that you do not read during your speech. You will bore your audience and lose their attention. You want to show the audience that you are familiar with your subject and you know your material. Using note cards as a guide will help you achieve this result. On your note cards, record information that is important and that will help you remember key points. For example, you may want to label one note card "Introduction," and on that card write down some sentences that will attract the audience's attention. Also, you will want to highlight your thesis statement on this note card. Make that the last sentence.

PRACTICE: CREATING NOTE CARDS

Create note cards for the speech you are giving to students at the community college about a career in policing. Remember to list the important points you will mention in your speech.

2. PREPARING FOR THE DELIVERY OF YOUR SPEECH

Once you have your note cards ready to go, your job in preparing for your speech is not over. You need to get yourself ready for that day. You may be familiar with your subject, but you want to be very familiar with the material in your speech so that you do not have to resort to lengthy pauses or stumble over your ideas when you are in front of the audience. Here are some tips to help you prepare for your speech so that it will go as smoothly as possible:

1. **Go over your material** as many times as you feel necessary to be comfortable with it, but do not memorize it. You want your audience to appreciate your familiarity with your subject and your material.

2. **Practise from your note cards.** Doing this will prepare you for using the reference material on the day of your speech. There will be no surprises because you will be familiar with the information on your note cards.

3. **Prepare any visual aids** that you will use in your speech. Visual aids are a must. They help keep the audience interested in what you have to say. Before you give your speech, have all of your visual aids ready. You may need to have a PowerPoint presentation ready, or you may want to get a clip of a movie, just at the right spot. Have these aids ready before you begin your speech.

4. **Practise using your visual aids.** You do not want to appear before your audience and find you do not know how to use the computer or you do not know how to work the DVD. Practise these things ahead of time so that everything runs smoothly on the day of your speech.

5. **Practise your speech using your visual aids.** Go through your entire production. Rehearse your speech and use your visual aids to ensure that everything fits together nicely.

6. **Gear yourself up so that you are a confident speaker.** You must prepare mentally before you give your speech. You want to appear confident when you are in front of an audience. Knowing your material is half the battle, but you must also create and maintain a positive attitude about yourself and your speech. Tell yourself you can do it and be confident. Keep this in mind. Show the audience your confidence.

PRACTICE: PREPARING FOR THE SPEECH

You are getting ready to deliver your speech to the law enforcement students. Prepare yourself for your speech. Follow the tips given for preparing yourself for the delivery of your speech and practise before it actually takes place.

KNOWING YOUR AUDIENCE

Before you prepare your speech, you must know who will be your audience. Different audiences call for different approaches. For example, if your topic is "Using Pepper Spray," your speech to a group of college students will differ from one you would give to a group of fellow police officers, who already have some knowledge of this subject. Your speech should be designed to build on your audience's previous knowledge of the subject. You do not want to present your audience with information they already know, and you do not want to start at such an advanced level that your audience is lost and finds it difficult to follow because of their lack of knowledge of the subject. If you are unsure about the group you will be speaking to and their knowledge of the subject, find out from the person who invited you to give the speech. This person will be able to give you an idea about the group's knowledge of your topic.

PRACTICE: KNOWING YOUR AUDIENCE

Evaluate your audience. What do you believe your audience will want to know about a career in policing? Use your classmates as a mock audience. Write down what you think they want to know about policing and what they might be interested in.

USING VISUAL AIDS

You must use visual aids to enhance your speech. They will keep the audience interested in what you have to say. Visual aids can take many forms, such as overhead transparencies, clips from a movie, handouts, the blackboard, PowerPoint presentations, and slides, and can be used to enhance your speech. However, you need to incorporate them properly to make your speech effective. Here are some tips to follow when preparing your aids to ensure they will aid your cause:

1. **Use visual aids that are large enough to be seen by everyone in the audience.** You don't want your audience to have to squint to see what you are presenting.

2. **Make all the visual aids neat.** Don't produce visual aids that are cluttered and hard to follow. Keep them neat and presentable.

3. **Make your visual aids interesting.** You want your speech to be memorable. Use aids that will keep the audience interested in what you have to say.

4. **Don't overuse visual aids.** Having the entire content of your speech on overhead transparencies or on PowerPoint slides and expecting your audience to follow along is not a good way to go. This will have an adverse effect on your audience. You will bore people and lose their attention.

5. **Refer to your visual aids throughout your speech.** Do not just put up a PowerPoint slide and then say nothing about it. You need to refer to it when you put it up.

6. **Ensure your equipment is in working order before giving your speech.** Try out the computer and the PowerPoint slides. You do not want to start your speech and then find out your machinery is not working.

PRACTICE: PREPARING A VISUAL AID

Prepare a visual aid to use for your speech to the law enforcement students. Follow the tips given in "Using Visual Aids."

3. DELIVERING YOUR SPEECH

The day has come for you to deliver your speech. You have practised, put your visual aids in place, and prepared yourself mentally. Now all you have to do is get up in front of the audience and present. Here are some tips to keep in mind:

1. **Get your visual aids and yourself in place.** Before actually getting into place, make sure your aids are all working and in order. Also, find that perfect spot for yourself. Do you want to stand behind a lectern? Do you want it out of your way? Make the changes you need so that you are comfortable.

2. **Begin without referring to your notes.** You don't want to start by looking at your note cards. You want to look at the audience and give them the impression that you know what you are talking about.

3. **Maintain eye contact with your audience.** This signals to your audience that you are a confident speaker and you are familiar with your subject. Look at your audience and express interest. Of course, you can refer to your notes, but make sure that you return your attention to your audience.

4. **Do not read!** This is an absolute must. Again, it is important that you be familiar with your material and that you refer to your notes only occasionally. You will lose your audience if you read, and you will create the impression that you are not familiar with your material.

5. **Avoid distracting mannerisms.** Do not stand in front of the audience and chew gum or twirl your hair. Portray confidence and stand with poise, using animated facial and hand gestures to keep your audience interested in what you have to say.

6. **Speak clearly and loudly.** Ensure that everyone can hear you, and speak clearly. Your audience will be distracted from the content of your speech if they have to strain to hear what you are saying.

7. **Dress appropriately.** Remember, you are speaking as a professional. As a police officer, it may be appropriate for you to wear your uniform.

8. **Speak with enthusiasm.** This is very important. Your audience will share your enthusiasm and become interested in your subject. Show the audience that this subject really means something to you and that it should be important to them, too.

PRACTICE: DELIVERING YOUR SPEECH

The time has come. You have prepared and are now ready to deliver your speech to the law enforcement students. Using your class as your audience, assume you are the constable giving the speech. To capture and maintain your audience's attention, use the tips given in "Delivering Your Speech."

AFTER YOU ARE FINISHED SPEAKING

Once you have finished speaking, you may sigh with relief. It is nearly over. Here are a few things you need to remember to do:

1. **Don't pack up right away.** Once you have finished making your point, don't assume your work is over. Don't start putting your materials away. This might give your audience the impression that you want to leave as soon as possible. You want to convey to your audience that you are available.

2. **Allow your audience the opportunity to ask questions.** Let your audience know that you would be pleased to answer any questions they might have on your subject. Ask the audience if they have any questions, and let them know you would be happy to answer them. Do not let your body language convey the message that you really don't want to answer questions or talk about your subject further.

3. **End appropriately.** Thank the audience for giving you the opportunity to speak to them. Let them know you would be happy to provide them with any further information that they might want on the subject.

4. EVALUATING YOUR SPEECH

Because your audience has gone does not mean that your work has ended. At this point, you should evaluate your speech. Remember, if you are called on to speak again, you will want to understand what went well, what you should use and do again, as well as what didn't go well and what you should avoid next time. It is important to evaluate yourself. Here are a few things you should consider after your speech:

1. **Did your visual aids work effectively?** Were the visual aids that you used received well by the audience? Did anyone have to squint in order to see your visuals, or did everyone seem to follow along easily? Did the visual aids fit well with the speech?

2. **Did you speak clearly and loudly?** Was everyone able to hear you without having to ask you to speak up? Also, try to remember whether you were asked to repeat something. If you were, this is an indication that, perhaps, at times, you did not speak clearly enough.

3. **Did the audience seem interested in what you had to say?** Did you keep the audience attentive and hooked on your subject?

4. **Did the audience ask questions?** This may be an indication to you that the audience found your speech interesting.

5. **Overall, what went well?** What was it about your speech that you enjoyed and what did the audience react to positively? Keep these things in mind.

6. **Overall, was there something you would change?** Was there something in your speech that you were not comfortable with or something your audience did not react to well? If so, keep these things in mind, and plan to change them.

PRACTICE: EVALUATING YOUR SPEECH

You are now ready to evaluate your speech. Using the guide given in "Evaluating Your Speech," assess how your speech went. What went well? What would you change? Write these things down.

DISCUSSION QUESTIONS

1. Give some examples of situations in which a police officer might be asked to give a speech.

2. What are the four stages in giving speeches? Briefly explain what is involved in each stage.

3. Sketch the format of the speech.

4. What are some tips for delivering an effective speech?

5. What must one remember when using visual aids?

ACTIVITY: DELIVERING A SPEECH

A. Assume your professor has asked you to give a speech to local high school students interested in getting into the law enforcement programs at your college. Prepare the speech, keeping in mind all the guidelines discussed. Prepare a visual aid for your speech. Give the speech to your classmates or invite interested high school students to hear you speak.

B. Select a topic from the list below, or choose your own topic related to policing. Research your topic, and prepare a 10-minute speech based on your findings. When you are preparing the speech, assume that you are a constable presenting to a specific group. Follow the tips listed below to prepare your speech.

1. Choose a topic.

2. Consider your audience.

3. Gather your research material.

4. Complete an outline.

5. Compose the written content.

6. Make note cards based on your written work.

7. Produce visual aids.

8. Practise your speech.

9. Deliver your speech, keeping in mind all the guidelines provided.

10. Evaluate your speech.

Possible Topics

1. Community Policing
2. Weapons in Policing
3. Canine Division
4. Policing Department Divisions
5. Forensics and Policing
6. Profile of a Police Officer
7. Policing Recruitment Process
8. Stress in Policing
9. History of Policing
10. Investigations

Chapter 7

Workshops

LEARNING OUTCOMES

Upon successful completion of this chapter, you should be able to

1. define a workshop
2. recognize the difference between a workshop and a speech
3. recognize and apply the three stages involved in conducting a workshop
4. recognize and apply the structure of the workshop
5. conduct an effective workshop
6. evaluate a workshop effectively

INTRODUCTION

As a police officer, there may be times you are called upon to conduct a workshop as opposed to giving a speech. There is a difference between giving a speech and conducting a workshop. Giving a speech involves presenting information to the participants. Your participants are listening to what you have to say and learning from what you tell them. With a speech, little or no interaction is required for learning to take place. A workshop involves interaction, because you are asking the participants to engage actively in the learning. This means that some hands-on or interactive exercises will occur.

For example, assume you are a sergeant in charge of training. You may conduct a pepper-spray workshop for constables in your division. After discussing the effects of pepper spray, officers may volunteer to be sprayed to experience the effects of this weapon, thereby learning by experimenting. You might also be asked to facilitate a workshop on interviewing skills. During this workshop, you might ask your officers to practise their interviewing skills with one another after you have given them the theory and techniques. There are other characteristics of a workshop that you should keep in mind.

- Workshops involve group work or individual practice.
- Workshops are short term. You may have as few as two hours to get your information across.

- Facilitators lead workshops. Facilitators are individuals with extensive knowledge or backgrounds related to the topic.
- Workshops usually revolve around some problem. For example, a sergeant may conduct a workshop on pepper spray because many constables have noted that they don't know how to use the weapon properly. Or a sergeant may provide a workshop on interviewing skills because it is apparent some officers need practice in this area.

Examine the following chart to see the differences between a speech and a workshop.

Speech	Workshop
✓ Involves passing on information	✓ Involves passing on information and having participants engage actively in learning
✓ Includes an introduction, body, and conclusion	✓ Includes an introduction, body, and conclusion
✓ Involves the speaker presenting information and participants learning through listening	✓ Involves the speaker presenting information and the participants learning through practice
✓ Involves little or no audience participation	✓ Always involves participant participation
✓ Presentation may not necessarily revolve around a particular problem	✓ Revolves around a particular problem
✓ May involve visual aids	✓ May involve visual aids

LEADING A WORKSHOP

As the leader of the workshop, there are several things you need to do. Review the list below to get a clear understanding of what is expected of you.

- Know your material and your agenda well.
- Make all the necessary arrangements in advance, such as preparing agendas, handouts, audiovisual presentations, room bookings, and room set-ups.
- While participants are involved in an activity, walk around the room and talk to individuals. The participants may have more questions or need further clarification.
- Keep track of the time. You want to keep on schedule and finish your workshop on time.
- Once the participants have completed their activity, ask them for feedback. As the leader, give the participants some feedback on what they accomplished.

STAGES IN CONDUCTING A WORKSHOP

There are three stages you should consider when conducting a workshop. These are

1. Planning the Workshop
2. Conducting the Workshop
3. Evaluating the Workshop

1. PLANNING THE WORKSHOP

In preparing for the workshop, there are several things you need to do. These include assessing the needs, specifying the earning objectives, designing your learning activities, making appropriate arrangements, and rehearsing.

Assessing the Needs

One of the first things you need to do is to determine why this workshop is necessary. Assessing the needs includes figuring out the problem, why it exists, and what should be done to overcome it. Once you know why you need to conduct the workshop, you can begin to organize it.

PRACTICE: PLANNING THE WORKSHOP

Assume you are a sergeant at the Metropolitan Police Department. Several of your constables are not familiar with the proper report-writing format and have requested training in this area. You have decided to approach this training as a workshop. Your first step in preparing this workshop is to assess the needs. Keeping in mind what was discussed about assessing needs, write a paragraph on what the problem is, why it exists, and how the constables can overcome the problem. You may need to fabricate some of the information for this exercise.

Specifying the Learning Objectives

Specifying the objectives stems from assessing the needs. Once you know why the workshop is necessary, you can then specify the objectives. Learning objectives are statements that indicate what you want your participants to learn.

PRACTICE: WRITING LEARNING OBJECTIVES

Once you have assessed the needs for the report-writing workshop, list the learning objectives for this workshop. What should your officers know by the end of the workshop?

Designing Learning Activities

Because the workshop is participatory, you need to think of activities to enhance your participants' learning. These learning activities may include games, puzzles, case scenarios, practising techniques, discussions, etc. When designing your learning activity, keep these things in mind:

- Think about your participants' capabilities and limitations. Choose an activity you know your participants can accomplish — don't make the activity too difficult for them, nor too easy.

- Choose an activity that will keep your participants interested. The activity should be both fun and valuable. Don't bore them.

- Choose an activity that is relevant to the topic and learning objectives. Remember you want your participants to learn.

- Keep the time frame in mind. Don't choose an activity that is going to take 45 minutes to complete when you only have an hour in total to work with; you will

```
INTERVIEWING SKILLS WORKSHOP

Room E456
Police Services Building
Thursday, June 15, 2006
1300–1500
Facilitator – Sergeant Erin Etak
Agenda
Introductions – 10 minutes
Purpose of Workshop – 10 minutes
Discussion of Different Interviewing Techniques – 30 minutes
Group Activity – Practising Techniques – 30 minutes
Group Activity Feedback – 15 minutes
Questions from Participants – 10 minutes
Conclude Workshop – 5 minutes
Workshop Evaluation – 10 minutes
```

have to rush through everything else. Many people design an agenda to pace their workshop and distribute it to the participants so they know exactly what will happen during the workshop and how long each item will take. See the sample agenda above.

- Keep the location of your workshop in mind. Will the location be appropriate for your activity? For example, if your activity involves group work, ensure the room has enough space to accommodate group work.

PRACTICE: DESIGNING AN ACTIVITY

Again, keeping in mind the report-writing workshop, design an activity to help your participants achieve the workshop's objective(s). For the purpose of this topic, let's assume that the officers will work individually.

PRACTICE: DESIGNING AN AGENDA

Design an agenda for the report-writing workshop. Let's assume the workshop will be two hours in length.

Making Appropriate Arrangements

Before you can put on the workshop, you need to make all the necessary arrangements. You will need to consider some of the following items:

- Book the appropriate room where you want the workshop to take place.

- Make arrangements to have all your material before the workshop takes place. If you need handouts prepared or photocopied, prepare them ahead of time. Don't leave these things to the last minute. You may not get things finished and it would then leave a bad impression with the participants.

- Inform your participants that the workshop will be taking place. You can do this in memo, email, or letter format — whichever is the most appropriate. Don't forget to give the participants the necessary details and information they will need, such as the *where*, *what*, *why*, *when*, and *who*.

PRACTICE: MAKING ARRANGEMENTS

In working to put together the report-writing workshop, assume you are ready to make all the appropriate arrangements. Make all the necessary arrangements, such as locating a room, writing a memo or email to let your officers know about the workshop, creating any handouts or overheads, and preparing an agenda for your participants.

Rehearsal

Just as you rehearse before giving a speech, so you should rehearse before conducting a workshop. Here are some questions to consider:

- Do you have a structured introduction, body, and conclusion to your workshop?
- Are your activities appropriate to the learning taking place?
- Are the activities geared to the audience?
- Are the handouts, overhead transparencies, and PowerPoint presentations ready?
- Is the location of the workshop appropriate for the proposed activity?

STRUCTURE OF THE WORKSHOP

Introduction

Background information related to the topic
Purpose of the workshop (This is your thesis statement.)

Body

Provide new information the participants will need to know related to the topic.
Lead into an activity.
Have the participants report on their activity and findings, if appropriate.

Conclusion

Restate your thesis (purpose) and important points of the workshop.
Close the workshop with closing comments on the topic and workshop itself.
Leave time for participants to ask questions.

---

2. CONDUCTING THE WORKSHOP

The day has finally come to put all of your planning into action. Your attention now shifts to the workshop itself. If you have planned well, made all the necessary arrangements, and rehearsed, you will be able to guide the workshop through to a successful conclusion. To ensure that everything runs smoothly, follow these tips:

- **Arrive at the location ahead of time.** This will give you time to arrange tables and chairs if necessary and to make sure your equipment is working properly. If you are early and discover the overhead projector or computer is not working properly, you will have time to get a replacement or make other arrangements.

- As mentioned earlier, **have all handouts and learning activities ready before the workshop**. You do not want to look unprofessional.

- **Create a positive learning climate.** As participants arrive, make them feel welcome and comfortable. Perhaps you can engage in small talk with them. Be friendly and approachable.

- **Be relaxed.** If you have made all the necessary arrangements and have rehearsed, everything will be fine. Try to relax and enjoy the experience.

- **Be professional.** Remember you are facilitating this workshop because you have expertise in this area, and you are going to share that expertise with a group of people. Even if you know the participants, approach the workshop professionally. Dress professionally and act accordingly.

- Throughout the workshop, **speak clearly.** Your participants want to hear everything you say, and understand everything you say without having to strain. Practise this in your rehearsal and follow through during the workshop.

- **Be enthusiastic.** Show your participants that you are excited about the material you are presenting. Your enthusiasm will inspire them.

- Once the workshop is complete, **leave time for questions.** Let your participants know that you welcome questions and are happy to answer them.

PRACTICE: CONDUCTING THE WORKSHOP

Based on the report-writing workshop, follow through and conduct the workshop for a group of students. Assume you are the sergeant facilitating the workshop, and that the participants are students acting as constables.

3. EVALUATING THE WORKSHOP

After the workshop is complete, do not immediately collect your materials and head out the door. An evaluation of the workshop is necessary. You can evaluate the workshop in two ways. First, as the facilitator, you should ask yourself a series of questions and write down the answers so that you can refer to them if you should conduct the workshop again. Ask yourself

1. Did the learning activity work effectively?

2. Did the audience seem interested in what I had to say and in what we were doing?

3. What went well?

4. Was there anything I would do differently next time?

5. Did I meet the learning objectives of the workshop?

Your participants should also have the opportunity to evaluate the workshop. Often participants are asked to fill in an evaluation form at the end of the workshop or to give their feedback about it. The evaluation will cover similar questions that the facilitator might ask himself or herself, to get an idea of how the participants felt about their experience. The feedback might be useful for any future workshops you conduct. If you are designing an evaluation form for the participants, consider these questions:

- Did the workshop meet your needs?

- Was the workshop easy to follow, and did it hold your interest?
- Did the activity fit well with the workshop and did it work effectively?
- Did you enjoy the presenter?
- What were the highlights of the workshop?
- Was there anything about the workshop that you think needs to be changed? If so, list the change(s) that you feel is (are) necessary.

Some evaluation forms are designed with a rating scale to make it easier for the participant to fill out. Here is an example:

Workshop Evaluation

1=not at all   2=somewhat   3=definitely

The workshop met my expectations.
1   2   3

I learned material from the workshop that I will be able to use in my professional career.
1   2   3

The workshop was interesting and held my interest.
1   2   3

The facilitator was knowledgeable and interesting.
1   2   3

Overall, I enjoyed the workshop.
1   2   3

Additional Comments

_____

_____

_____

_____

_____

_____

_____

_____

_____

_____

PRACTICE: CONDUCTING THE EVALUATION

Design an evaluation form for you as the facilitator to complete, and an evaluation form for the participants of the workshop to complete. After you have completed the workshop on report writing, have the participants fill in the evaluation form. Then complete your own form. Look at the completed evaluations and write a summary of the results.

DISCUSSION QUESTIONS

1. Define *workshop*. How does a workshop differ from a speech?
2. Discuss the steps involved in planning a workshop.
3. Discuss some of the tips involved in conducting a workshop.
4. How would you evaluate a workshop?
5. List some possible workshops that police officers might facilitate.

ACTIVITY: FACILITATING A WORKSHOP

Select a topic from the list below, or choose a topic of your own related to policing. Research your topic and prepare a workshop. When you are preparing and conducting your workshop, assume you are a police officer presenting to a specific group. Follow the tips below as you prepare:

1. Assess the needs: Why is this workshop necessary?
2. Write learning objectives: What do you hope the participants will gain from the workshop?
3. Design a learning activity appropriate to the workshop.
4. Design an agenda for the workshop.
5. Make appropriate arrangements. Don't forget to book a room and send the participants a memo, letter, or email informing them of the workshop. Prepare all material and handouts.
6. Rehearse before you conduct the workshop.
7. Conduct the workshop.
8. Evaluate the workshop.

Possible Topics

1. Interviewing Skills
2. Working with a Canine
3. Conducting an Investigation
4. Proper Use of Handcuffs
5. Keeping in Shape
6. Proper Arrest Procedures
7. Note-Taking Procedures
8. Dealing with a Crisis Situation
9. Proper Weapon Handling
10. Coping with a High-Stress Job

Chapter 8

Sentence Skills

LEARNING OUTCOMES

Upon successful completion of this chapter, you will be able to

1. recognize and correctly spell frequently misspelled words in police reports.
2. recognize and correctly use homonyms commonly used in police writing.
3. correctly apply subject–verb agreement
4. correctly apply pronoun agreement, reference, and point of view
5. correctly apply basic grammatical rules for correcting sentence fragments and run-ons
6. correctly apply basic grammatical rules in relation to parallel structure and proper modification
7. correctly apply comma rules

WORDS OFTEN MISSPELLED IN POLICE REPORTS

Here is a list of words often misspelled in police reports. Look over the list and be aware of the common errors that are made.

abduction	benefited	height	precinct	scene
accelerated	burglary	homicide	premises	sergeant
accessories	coercion	interrogate	prejudice	strangulation
accident	complainant	intoxication	prevalent	subpoena
acquire	conscious	investigation	principal	suicide
acquitted	controversial	its (it's)	principle	surrender
affidavit	conviction	juvenile	privilege	surveillance
altercation	corpse	larceny	procedure	suspicion
apparatus	counterfeit	occurrence	prostitution	testimony
arguing	criminal	offence	pursue	than (then)
argument	defendant	official	pursuit	there (they're)
arson	embezzlement	patrolling	referring	(their)
assault	fraudulent	possession	sabotage	thorough

belief	environment	pedestrian	resistance	to/too/two
believe	extortion	penalize	rhythm	villain
beneficial	forcible	personnel	robbery	warrant

ACTIVITY

Choose a partner. Take turns being the dictator and the speller. Keep track of how many words you spell correctly. If there are words that you have problems with, keep track of them by writing them in the space provided below. Continue to practise them until you can spell them correctly.

_____

_____

_____

_____

_____

HOMONYMS

Homonyms are words that sound alike but have different meanings and are spelled differently. Look over the list. You should learn the difference and use the correctly spelled word.

beat/beet	knew/new	prey/pray	steal/steel
board/bored	knot/not	presence/presents	their/there/they're
break/brake	know/no	principal/principle	to/too/two
buy/by/bye	lie/lye	rap/wrap	waist/waste
corps/corpse	made/maid	right/write	weigh/way
ceiling/sealing	maybe/may be	scene/seen	wear/where
cent/sent/scent	oar/or/ore	sense/cents	weak/week
cereal/serial	pain/pane	stairs/stares	
cite/sight/site	peace/piece	stake/steak	
its/it's	plain/plane	stationary/stationery	

ACTIVITY

Read each sentence and decide whether the sentence contains the wrong homonym. Rewrite the sentence and make the correction.

1. The suspect was wanted for stealing the money from the cash register.
2. The constables were dispatched to a brake and enter call and arrived at the seen within too minutes.

3. The constable likes to rite his notes at the seen of a crime.

4. Last weak, the suspect told the constable that she went against her principals and through the chair at the victim and than beet her.

5. The judge waisted a great deal of time on the case and did knot no that the next case was too be herd.

6. The detective noticed blood splattered on the sealing.

7. The police arrived on cite an our after the incident occurred.

8. The police dog followed the cent of the missing child.

9. A pear of shoes was found on the river bank.

10. Everyone was relieved when the serial killer was located.

SUBJECT-VERB AGREEMENT

One of the most common grammatical problems is failure to make the subject and verb in a sentence agree with each other. Here is the rule for subject–verb agreement.

> *Singular subjects take singular verbs.*
> *Plural subjects take plural verbs.*

Remember that singular words concern one person or thing. Plural words concern more than one person or thing.

Examples:

The police officer writes his reports immediately after the incidents occur. (singular subject and singular verb)

The police officers write their reports immediately after the incidents occur. (plural subject and plural verb)

Constable Tracey plays hockey on his days off. (singular subject and singular verb)

Constable Tracey and Constable Merrill are partners. (plural subject and plural verb)

There are tricky situations that can complicate subject–verb agreement. Look at the following situations.

1. **When words come between the subject and the verb**

One of the witnesses was able to get a good look at the suspect.

Who or what is the subject? *One*

The prepositional phrase *of the witnesses* does not contain the subject. Therefore, *One* is the subject. You will never find the subject in a prepositional phrase. Cross out the prepositional phrase, and you will then be able to identify your subject. If your subject is singular, use a singular verb. If your subject is plural, use a plural verb. In this case, because the subject is singular, a singular verb (*was*) is used. Here is a list of common prepositions that will create prepositional phrases. If you can't decide whether a sentence requires a singular or plural verb, check for prepositional phrases — they will help you locate your subject. Your subject will then become clearer.

about	before	by	inside	over
above	behind	during	into	through
across	below	except	of	to

among	beneath	for	off	towards
around	beside	from	on	under
at	between	in	onto	with

2. Compound subjects

Good listening and note-taking skills are required to write a police report.

What is the subject? *listening and note-taking skills*

Is the subject singular or plural? *plural*

Is the verb plural to match the subject? Yes, the verb is *are*

To create agreement in your sentence use a plural verb with a compound subject.

3. Verb before the subject

There were three break and enters that occurred last night.

What is the subject? *break and enters*

Is the subject singular or plural? *plural*

Is the verb plural to match the subject? Yes, *were* is the verb.

4. Indefinite pronouns

The following words, known as indefinite pronouns, always take singular verbs.

one	nobody	nothing	each
anyone	anybody	anything	either
everyone	everybody	everything	neither
someone	somebody	something	

Examples:

One of the constables is attending the computer workshop.

Anybody is welcomed to come and support the constables in their marathon.

Everything in the precinct is under control.

Either Constable Markoo or Constable Dupuis was dispatched to that call.

PRACTICE: SUBJECT-VERB AGREEMENT

Identify the correct verb for each sentence, ensuring that the subject agrees with the verb.

1. There (is, are) many issues to consider before the staff sergeant makes a decision.
2. Constable Phillips and Constable Fredovich (sings, sing) at the concert annually.
3. The patrol cars (is, are) undergoing major repair.
4. One of the officers (has, have) refused to complete the task.
5. Several of the officers in our department (has, have) been congratulated on a job well done.
6. Either Sergeant Millar or Sergeant Khan (is, are) going to present the department's concerns.

7. Someone from the department (is, are) going to be invited to the conference.

8. There (is, are) many officers who are pleased with the decision.

9. The witness and the victim (testify, testifies) at the trial.

10. There (is, are) three witnesses in the case.

PRONOUN AGREEMENT, REFERENCE, AND POINT OF VIEW

Pronoun Agreement

Antecedents

The basic rule to remember is this:

> *A pronoun must agree with its antecedent.*

A pronoun takes the place of a noun.

An *antecedent* is the noun that a pronoun replaces or refers to.

Examples:

The witness gave his statement. (*His* is a pronoun that refers to *witness*.

Witness is the antecedent.)

The police officers gave their opinions on the matter. (*Their* is a pronoun referring to *police officers*. *Police officers* is the antecedent.)

The victim was terrified that the suspect would come after her again. (*Her* is a pronoun referring to *victim*. *Victim* is the antecedent.)

A pronoun must agree with its antecedent. If the word your pronoun is replacing is singular, then your pronoun must be singular. If the word your pronoun is replacing is plural, then your pronoun must be plural.

Indefinite Pronouns

You have already looked at a chart of indefinite pronouns in the section on subject–verb agreement. You learned that when these words are used as subjects, they are singular and take singular verbs. So it makes sense that the pronouns that stand for or refer to them must be singular.

Examples:

Everyone in the force must do *his/her* best while on duty.

Each of the police officers will contribute *his/her* part to the program.

No one can truly say that in *his/her* mind *he/she* believes the accused is guilty.

In all the examples above, each pronoun is singular because it refers to an indefinite pronoun.

Pronoun Reference

Here is the rule:

> *A pronoun must clearly refer to the correct antecedent.*

Mistakes may occur if you fail to follow this rule.

Look at these examples.

Sergeant Vincent told Chief Thiessen that his plan didn't work.

Whose plan didn't work? The sergeant's or the chief's?

The victim pointed to the witness and said that he had saved his life.

Who saved whom?

If the police dog won't eat its food, feed it to the other police dog.

Feed what to the other police dog? The *food* or the *dog*?

You can only guess at the meaning of these sentences because you don't know to whom the pronouns refer. Look carefully at how the sentences have been corrected.

Sergeant Vincent told Chief Thiessen, "Your plan didn't work."

The victim pointed to the witness and said, "He saved my life."

If the police dog won't eat its food, feed the food to the other police dog.

Pronoun Point of View

There are three categories of *person* that you can use when you write or speak. They are:

Person	Singular	Plural
First Person	I	We
Second Person	You	You
Third Person	She, He, It Words that end with "one" Words that end with "body" Words that end with "thing"	They

Here is the rule to remember:

> *Do not mix persons unless the meaning requires it.*

You must be consistent. If you begin a discussion in the second person, you must use second person throughout.

Look at this error:

If *you* wish to become a police officer, *one* must be fit.

Be consistent and use the same person throughout the sentence.

Correction:

If *you* wish to become a police officer, *you* must be fit.

or

If *one* wishes to become a police officer, *one* must be fit.

Here is another example:

When *we* asked about training, *we* learned that *you* could take only one first-aid course.

Correction:

When *we* asked about training, *we* learned that *we* could take only one first-aid course.

PRACTICE: PRONOUN AGREEMENT, REFERENCE, AND POINT OF VIEW

In the following sentences, correct any errors in pronoun agreement, reference, or point of view.

1. Constable Livingstone looked at the victim and ^the victim said that he had feared for his life.

2. After twenty minutes, every one of the police officers was looking at their watches and thinking the presentation would never end.

3. Constable Siddiqui and Constable Bimson forgot ~~his~~ their notebook at the precinct.

4. Someone should lend ~~their~~ his/her copy of the report to the constables.

5. The constable backed his cruiser into a mail box and dented it.

6. The victim felt that the accused should have been more careful with her bike when she lent it to her because she was a good friend of her sister.

7. In many police departments across Ontario, they use the computer system.

8. I expect every officer to do their best at all times.

9. If you do not exercise daily, one will not be ready for the physical testing.

10. One will become a police officer if you make it a personal goal.

SENTENCE FRAGMENTS

Any group of words that is punctuated as a sentence but does not have a subject or a complete verb or does not express a complete thought is a sentence fragment.

Here are a few examples:

Was attacked by her neighbour.

Who was attacked? The sentence does not tell you. The subject is missing.

The accused attempting to get away.

Part of the verb is missing. Remember that a verb ending in *-ing* must have a helping verb in front of it.

For prisoners in Ontario.

Both the subject and the verb are missing.

Regarding the Mathewson case from last week.

Both the subject and verb are missing.

What do you do with the fragments?

> *To change a sentence fragment into a complete sentence,*
> *add whatever is missing: a subject, a verb, or both.*

You may need to add a *subject*.

The victim was attacked by her neighbour.

You may need to add part of a *verb*.

The accused was attempting to get away.

You may need to add *both a subject and a verb*.

Physical activity will be provided for prisoners in Ontario.

The Crown and defence attorneys will meet regarding the Mathewson case.

Also, keep in mind dependent and independent clauses. An independent clause is one that makes complete sense on its own. It can stand alone as a sentence. A dependent clause cannot stand alone as a sentence; it depends on another clause to make sense.

A dependent clause must be attached to an independent clause. If it stands alone, it is a sentence fragment.

Here is a dependent clause:

Because he was convicted. The accused would spend 15 years in prison.

Because he was convicted is a dependent clause. It is a fragment and cannot stand alone.

Here is how to correct it:

Because he was convicted, the accused would spend 15 years in prison.

PRACTICE: SENTENCE FRAGMENTS

Correct any sentence fragments in the following exercises.

1. The suspect 1.83 m tall.
2. The police officer was horrified. Because the suspect showed no remorse.
3. The victim crying when she discovered that her house was ransacked.
4. Testifying in court.
5. Listening to the witness. The officer nodded to show concern.
6. After the incident.
7. The police officer sounding his siren.
8. The sergeant announced that there would be a meeting. To discuss the changes to the policy.
9. For the witnesses at the trial.
10. The jurors made a decision. Based on the evidence provided at the trial.

RUN-ON SENTENCES

Any group of words that has a subject and a verb and expresses a complete thought is a sentence and should be punctuated as a complete sentence. Combining two or more complete sentences into one without the proper punctuation creates a run-on sentence. There are two types of run-on sentences. Two independent clauses written together with no punctuation are a fused sentence. Two independent clauses joined by only a comma are known as a comma splice.

Here are some examples:

Constable Michaels warned the motorist about the dangers of speeding,Constable Levitt wrote the speeding ticket.

Here we have two complete sentences joined together with no punctuation at all. A fused sentence has been created.

Constable Picard decided she would take the weapons training, Constable Pino decided he didn't need it.

Here a comma joins two complete sentences. A comma splice has been created because a comma has been used incorrectly to join two independent clauses.

There are three ways to correct a run-on sentence.

> *1. Make the two independent clauses into two sentences using a period.*

Example: Constable Michaels warned the motorist about the dangers of speeding. Constable Levitt wrote the speeding ticket.

> *2. Use a semicolon to join two independent clauses.*

Example: Constable Picard decided she would take the weapons training; Constable Pino decided he didn't need it.

> *3. Connect the two independent clauses with a comma and one of the following conjunctions:* for, and, nor, but, or, yet, so.

Example: Constable Picard decided she would take the weapons training, but Constable Pino decided he didn't need it.

Here is another example:

The staff sergeant noticed that a lot of crime was occurring in the neighbourhood she decided to have her officers patrol it more closely.

The staff sergeant noticed that a lot of crime was occurring in the neighbourhood is a complete sentence.

She decided to have her officers patrol it more closely is a complete sentence.

This run-on sentence can be corrected using any one of the three methods cited above.

1. The staff sergeant noticed that a lot of crime was occurring in the neighbourhood. She decided to have her officers patrol it more closely.
2. The staff sergeant noticed that a lot of crime was occurring in the neighbourhood; she decided to have her officers patrol it more closely.
3. The staff sergeant noticed that a lot of crime was occurring in the neighbourhood, so she decided to have her officers patrol it more closely.

TRANSITIONAL WORDS

Transitional words may be used to join two complete sentences. If you are using a transitional word to join two sentences, use a semicolon before the transitional word and a comma after it.

Here is a list of transitional words and phrases:

therefore	otherwise	nevertheless	consequently	however	furthermore
moreover	meanwhile	as a result	thus	also	on the other hand

Here are some examples:

The officers decided they had better back down; otherwise, they believed they would put the victim's life in jeopardy.

The suspect was arrested; as a result, the victim was now feeling at ease.

The case was solved; however, the accused still needed to stand trial.

Keep in mind that a semicolon and a comma surround the transitional word only when it joins two sentences. Note the following example.

The witness should not assume, therefore, that he would not have to testify.

These are not complete sentences; they are two dependent clauses. Do not use a semicolon in this example.

PRACTICE: RUN-ON SENTENCES

Correct the following run-on sentences.

1. The sergeant retired from the force he continued to coach the baseball team.
2. The constable knew something was wrong, the house was much too quiet.
3. The officer was dispatched to a break and enter, therefore, she knew she would need her notebook.
4. The constable reported the incident to his sergeant he knew the case needed further investigation.
5. Most of the investigations were complete, however, some of the evidence was still missing.
6. All officers were encouraged to attend the seminar they all needed practice in writing police reports.
7. The victim called the police immediately she knew the suspect was getting away.
8. The witness told the police he wanted nothing to do with the case, he feared for his safety.
9. The neighbours had a meeting, the officer explained how they could protect themselves from future break and enters.
10. The officer consoled the victim, he knew that the victim was upset.

PARALLEL STRUCTURE

When writing about items in a series, you must be sure that all the items are parallel — avoid *faulty parallelism*. The items must be written in the same grammatical form.

Look at this example:

Police report writing involves listening to the victim, taking notes, and to write the narrative.

The items in this sentence are not parallel. Two items end in *-ing*, but the third item (*to write*) does not. To make the sentence grammatically correct, you must make all the items in the list take the same form.

Here is the correction:

Police report writing involves listening to the victim, taking notes, and writing the narrative.

Here are other examples that compare parallel and not parallel sentences.

1. **Not Parallel**
    Constable Milward is kind, considerate, and likes to help.

    Parallel
    Constable Milward is kind, considerate, and helpful.

2. **Not Parallel**
    The victim bit, kicked, and was punching the attacker.

    Parallel
    The victim bit, kicked, and punched the attacker.

3. **Not Parallel**
    The witness told the officer what he saw, heard, and was finding at the scene of the crime.

    Parallel
    The witness told the officer what he saw, heard, and found at the scene of the crime.

PRACTICE: PARALLEL STRUCTURE

Rewrite the following sentences using parallel form.

1. Eating his lunch, reading the paper, and to take a nap were things the witness liked to do on his lunch break.
2. The staff sergeant called the meeting; the agenda was prepared by the typist.
3. The chief read over the report, made some comments, and had it sent back to the officer for revision.
4. Over the last year the prisoner has had time to take courses in English, think about the crime he committed, and he had time to think about his future.
5. The officers were advised to patrol the area, watch for anyone suspicious, and to call for back up if needed.
6. The suspect jumped the fence to get away from the police; to capture the suspect, the police went after him.
7. The suspect accused the officer of making up evidence, discriminating against him, and of physically hurting him.
8. The witness watched the suspect out of his window, took down the licence plate number, and will call the police.

9. Taking down the licence number, phoning the police, and to testify in court were all responsibilities of the witness.

10. The victim exhibited self-destructive behaviour; a phone call was made by the police officer to the assault centre.

MISPLACED AND DANGLING MODIFIERS

Modifiers are used to describe, explain, or limit another word in a sentence. Modifiers give additional information about another word in the sentence. Problems occur when modifiers are used incorrectly; they can cause confusion and amusement. You need to recognize and correct two kinds of modifier problems. These are misplaced modifiers and dangling modifiers.

Misplaced Modifiers

You should put the modifier as close as possible to the word it is describing. If you fail to do this, you will change the meaning of the sentence.

Look at these examples:

1. Constable Murdock told the suspect only what he needed to.
   As written, the sentence indicates that Constable Murdock didn't tell the suspect anything else. *Only* is the modifier and it modifies *what*.

2. Constable Murdock told *only* the suspect what he needed to.
   As written, the sentence indicates that Constable Murdock told the suspect and no one else. *Only* is modifying *suspect* in this case.

3. *Only* Constable Murdock told the suspect what he needed to.
   As written, the sentence indicates that it was only Constable Murdock and no one else who told the suspect. *Only* is modifying *Constable Murdock* in this case.

Here is the rule to remember:

> *Always place the modifier next to the word it is describing.*

Look at this example.

The suspect carried the stereo with arthritis.

The sentence is written as if the stereo has the arthritis. This is not what the writer means. It is the suspect who has arthritis. To correct the sentence, the writer needs to place the modifier *with arthritis* next to the word it is describing, which is *suspect*.

The correction would read

With arthritis, the suspect carried the stereo.

Here is another example.

The witness looked into the neighbour's window with binoculars.

In this example, a reader can interpret the sentence to mean that the window has binoculars. This is not the writer's intention. It is the witness who had the binoculars, not the window.

The correction would read

With binoculars, the witness looked into the neighbour's window.

Dangling Modifiers

Dangling modifiers usually occur when the modifying phrase opens the sentence and does not appear close to the word it is describing. Look at this example.

> After 25 years of hard work, the department awarded Sergeant Evans with a party.

This sentence seems to say that the department worked for 25 years. What the writer really means is that Sergeant Evans worked for 25 years. The writer needs to put the modifying phrase in the appropriate place.

Here is the correction.

> After 25 years of hard work, Sergeant Evans was rewarded with a party put on by the department.

Here is another example.

> Walking along the boardwalk, a motorcycle swerved and hit the victim.

This sentence would lead the reader to believe that the motorcycle was walking along the boardwalk. What the writer really wants it to say is that the victim was walking along the boardwalk. To correct this sentence the writer must put the modifying phrase close to the word it is describing. Here is the correction:

> Walking along the boardwalk, the victim was hit by a swerving motorcycle.

PRACTICE: MISPLACED AND DANGLING MODIFIERS

Correct any misplaced or dangling modifiers in the following sentences.

1. Running quickly into the bush, we watched the suspects get away.
2. Hanging on the office wall, Sergeant Wallis was proud of her certificate.
3. The burglar asked the victim to hand her the money with greed.
4. The witness told the constable everything she saw with tears in her eyes.
5. After his car accident, the constable told the victim his car was being transported to the garage.
6. The phone at the detachment almost rang 50 times in two minutes.
7. With great pride, the accused was arrested by the rookie officer.
8. The victim got away from the kidnapper with determination.
9. Constable Culgin arrested the accused with a grin.
10. After his getaway, the police officer thought about the suspect's whereabouts.

COMMAS

Commas are often misused in writing. They have six main uses. If you apply these six rules when you are writing, commas can become easier to use.

Here are the six comma rules, followed by examples:

> *1. Use a comma between items in a series.*

The burglar took the TV, VCR, stereo, and jewellery from the person's home.
The victim told the officer what happened, who he suspected, and what needed to be done.

> **2. Use a comma after an introductory phrase.**

When the victim called the police, she was sure they would be there within minutes.
Although the burglar got away, the police had a good idea of where he went.
Because the jury reached a verdict, the court was back in session.

> **3. Use commas around words that interrupt the flow of thought.**

The victim, who was hurt badly, was rushed to the hospital.
Constable Keeler, one of the force's best officers, was honoured at the banquet.
Detective Gray, who worked endlessly on the case, finally had a lead.

> **4. Place a comma between independent clauses when they are joined by the following conjunctions: for, and, nor, but, or, yet, so.**

Staff Sergeant Hodges was not happy about what had occurred, for many of his officers were injured during the incident.
The suspect wore a mask when he robbed the convenience store, and the clerk could not identify him in the line-up.
The witness did not get a good look at the kidnapper, but she did note the licence plate number of the car he was driving.

> **5. Use a comma with a direct quotation.**

The officer said, "Put the gun down and come forward."
"Don't make a sound," said the kidnapper. "You don't want to get hurt."
The witness said, "She was approximately six feet tall and had long black hair."

> **6. Use commas with dates.**

July 20, 2006, is when the incident happened.
The court date is set for August 4, 2006.

PRACTICE: COMMAS

Insert commas where necessary in each sentence.

1. Detective Anderson who is quite good at solving cases has decided to retire.
2. After much consideration the witness realized that her testimony was essential.
3. The jury had come to a decision and everyone in the courtroom was anxious to hear the outcome.
4. The suspect surrendered for she knew the police would catch her sooner or later.
5. "You must confess" insisted the detective.
6. The officer had his gun baton and pepper spray with him at all times.
7. One of his greatest strengths courage was now beginning to fade.
8. Because the robber had a knife the clerk did everything he asked.
9. Wasn't the customer afraid asked the reporter when the robber came after her with a knife?
10. There was a blizzard and the officers could not see anything.

DISCUSSION QUESTIONS

1. What is a sentence fragment? How can one correct a fragment?
2. What is a run-on sentence? How can one correct a run-on?
3. List the indefinite pronouns.
4. What rule must one remember about modifiers?
5. What is meant by parallel structure?

ACTIVITY: IDENTIFYING AND CORRECTING ERRORS

Identify the errors in each of the following sentences, and make the appropriate corrections. If a sentence is correct the way it is, leave it alone. You may find errors in pronoun use, subject–verb agreement, run-ons, fragments, misplaced modifiers, or parallel construction.

1. The officer stopped her car she wrote the motorist a speeding ticket.

_____

_____

2. The officer told the witness that he was lucky to be alive.

_____

_____

3. The victim's belongings was located in the suspect's car.

_____

_____

4. The suspect ran out the door into the van and drove off at high speed.

_____

_____

5. After making the arrest, the suspect was taken to jail by the constable.

_____

_____

6. Mary Jackson one of the witnesses in the case was relieved to hear the suspect was caught.

_____

_____

7. One of the officers learned their partners had been shot.

_____

_____

8. Each of the officers have their own notebook.

_____

_____

9. The bank robber told his partner that his plan did not work.

_____

_____

10. When you arrest a suspect, one should ensure that his rights are read.

_____

_____

11. The suspect with black hair, brown eyes.

_____

_____

12. The constable made the arrest and the suspect was put in prison.

_____

_____

13. Cried all night.

_____

_____

14. The inmate refused to eat his dinner, his appetite was lost after the incident had happened.

_____

_____

15. On her time off Sergeant Bookman enjoyed playing soccer, going camping, and to dance.

_____

_____

16. Because the victim was unconscious the officer called an ambulance.

_____

_____

17. Each of the officers were proud of their accomplishments.

_____

_____

18. The prisoner was not going to do as he was told, he felt he should not have to follow rules.

_____

_____

19. Jeff was studying law enforcement in college he knew he would make a great police officer.

_____

_____

20. Thinking it was the best decision.

_____

_____

21. It is important that the officers know how to write reports. The reports will be read by many lawyers.

_____

_____

22. The facility were minimum security.

_____

_____

23. It took the officer a while to review his notes, to dictate his report and check for any errors he may have made.

_____

_____

24. The constables who were the first ones at the incident were astonished by the damage that occurred.

_____

_____

25. Plead guilty or not guilty.

_____

_____

ACTIVITY: WRITING SENTENCES

Write a sentence for each item listed below.
1. Use a series showing parallelism.
2. Use "each" as the subject of the sentence.
3. Use a transitional word to join two complete sentences.
4. Use a semicolon to join two complete sentences.
5. Show the correct use of a pronoun reference.
6. Use three different comma rules.
7. Use a prepositional phrase in a sentence.
8. Use the second person.
9. Use a modifier in two different sentences.
10. Use the third person.

ACTIVITY

In the following police reports, you will find many of the errors discussed in this chapter. Rewrite the police report, correcting the errors.

Scenario One

On Monday, February 16, 2006, Constable Levine and Constable Markham was dispatched to a Brake and Enter call at 199 First Avenue at 1130 hours.

Constable Levine interviewed Monica Steffanelli who told Constable Levine the following. She left the house at 0800 hours and the doors and windows are secured. Upon returning home. Mrs. Steffanelli noticed the front door is open. When she enters the house. She notices that her tv, dvd player and staireo are missing from the living room. One of the Royal Dalton collectibles were taken from the cabinet. Neither of the prints were taken from the wall. Constable Levine examined the entry point, the livingroom and than went around the rest of the house, nothing else was disturbed.

Constable Levine tells Mrs. Steffanelli to call her insurance company. Constable Levine and I leave at 1230 hours.

Scenario Two

On Friday, June 22, 2006 at 1030 hours Constable Phillipe and Constable Oh were dispatched to 34 Magnum Street to take a report on a noise problem.

Constable Phillipe and Constable Oh were greeted by the complaintent Mr. Lou Valero. Mr. Valero tells the constables the following. His neighbours, Wesley Bucannen and Victoria Bucannen always have parties at there house. The parties last usually until too o'clock in the morning. Mr. Valero works at the steal factory and wakes up at 3 a.m.everyday for his 4 a.m. shift, Mr. Valero does knot get much sleep because of all the noise his neighbours make. Mr. Valero has often talked to his neighbours. About all the noise they make at the parties. He heres the people singing, dancing, and they seem to talk loudly. The neighbours Mr. And Mrs. Bucannen do knot seam to care. They continue to have these parties Mr. Valero felt he had know choice but to call the police on his neighbours.

Constable Phillipe assures Mr. Valero that he will talk to them. Constable Phillipe and Constable Oh leave the house at 1100 hours.

APPENDIX

SAMPLE POLICE TEMPLATES

The following police report templates have been provided courtesy of the Sault Ste. Marie Police Service and the Ontario Police Technology & Information Co-operative (OPTIC). This Appendix contains two copies of each form; to download additional copies, please visit the text web site at www.pearsoned.ca/text/turpin.

SAULT STE. MARIE POLICE SERVICE
DATA ENTRY GUIDE

FRAUDULENT DOCUMENT REPORT

OCCURRENCE #:	**TASK #:**
AUTHOR: (DICTATING OFFICER)	**REPORT TIME:**
ENTERED BY: (TRANSCRIBER)	**ENTERED TIME: (DEFAULTS TO PRESENT TIME)**
REMARKS:	
VICTIM NAME:	**VICTIM TYPE:**

REASON FOR REPORT:

☐	ACCOUNT CLOSED	☐	ENDORSEMENT FORGED	☐	SIGNATURE FORGED
☐	AMOUNT RAISED	☐	NO ACCOUNT	☐	STOLEN
☐	CERTIFICATE FORGED	☐	NON-SUFFICIENT FUNDS	☐	OTHER (CLARIFY)

DOCUMENT TYPE:

☐	COMPANY CHEQUE	☐	INVOICE	☐	SALES DRAFT
☐	COUNTER CHEQUE	☐	MONEY ORDER	☐	TRAVELLER'S CHEQUE
☐	CREDIT CARD	☐	PAYROLL CHEQUE	☐	OTHER (CLARIFY)
☐	GOVERNMENT CHEQUE	☐	PERSONAL CHEQUE		

DATE:	
ISSUED BY: (BANK)	**BRANCH:**
ACCT/DOC #:	**DOC SEQ #:**
PAYABLE TO:	**SIGNED BY:**
VALUE:	**CASH REC'D:**
WRITTEN BY COMPLAINANT:	**WRITTEN BY SUSPECT:**

IDENTIFICATION USED:

DRIVERS LICENCE #:		**PROVINCE:**	
VEHICLE LICENCE #:		**VEHICLE PROVINCE:**	
MAKE:	**MODEL:**		**COLOR:**
CARD TYPE:		**ISSUED BY:**	
CARD #:		**TEL:**	

NARRATIVE:

(TRANSCRIBE FRAUDULENT DOCUMENT REPORT NARRATIVE AS DICTATED)

SAULT STE. MARIE POLICE SERVICE
DATA ENTRY GUIDE

FRAUDULENT DOCUMENT REPORT

OCCURRENCE #:	TASK #:
AUTHOR: (DICTATING OFFICER)	REPORT TIME:
ENTERED BY: (TRANSCRIBER)	ENTERED TIME: (DEFAULTS TO PRESENT TIME)
REMARKS:	
VICTIM NAME:	VICTIM TYPE:

REASON FOR REPORT:

☐	ACCOUNT CLOSED	☐	ENDORSEMENT FORGED	☐	SIGNATURE FORGED
☐	AMOUNT RAISED	☐	NO ACCOUNT	☐	STOLEN
☐	CERTIFICATE FORGED	☐	NON-SUFFICIENT FUNDS	☐	OTHER (CLARIFY)

DOCUMENT TYPE:

☐	COMPANY CHEQUE	☐	INVOICE	☐	SALES DRAFT
☐	COUNTER CHEQUE	☐	MONEY ORDER	☐	TRAVELLER'S CHEQUE
☐	CREDIT CARD	☐	PAYROLL CHEQUE	☐	OTHER (CLARIFY)
☐	GOVERNMENT CHEQUE	☐	PERSONAL CHEQUE		

DATE:	
ISSUED BY: (BANK)	BRANCH:
ACCT/DOC #:	DOC SEQ #:
PAYABLE TO:	SIGNED BY:
VALUE:	CASH REC'D:
WRITTEN BY COMPLAINANT:	WRITTEN BY SUSPECT:

IDENTIFICATION USED:

DRIVERS LICENCE #:		PROVINCE:	
VEHICLE LICENCE #:		VEHICLE PROVINCE:	
MAKE:	MODEL:		COLOR:
CARD TYPE:		ISSUED BY:	
CARD #:		TEL:	

NARRATIVE:

(TRANSCRIBE FRAUDULENT DOCUMENT REPORT NARRATIVE AS DICTATED)

SAULT STE. MARIE POLICE SERVICE
DATA ENTRY GUIDE

WEAPON CAUSING INJURY

☐ NOT APPLICABLE	☐ UNKNOWN	☐ FULLY AUTOMATIC
☐ SAWED-OFF RIFLE OR SHOTGUN	☐ HANDGUN	☐ RIFLE OR SHOTGUN
☐ OTHER FIREARM	☐ KNIFE	☐ OTHER PIERCING OR CUTTING
☐ CLUB	☐ EXPLOSIVES	☐ FIRE
☐ PHYSICAL FORCE	☐ OTHER WEAPON	

LEVEL OF INJURY

☐ NOT APPLICABLE	☐ UNKNOWN	☐ NO INJURIES
☐ MINOR INJURY	☐ MAJOR PHYSICAL INJURY	☐ LOSS OF LIFE

ACCUSED IS (RELATIONSHIP TO VICTIM)

☐ UNKNOWN	☐ SPOUSE	☐ SEP. OR DIVORCED
☐ PARENT	☐ CHILD	☐ OTHER IMMEDIATE
☐ EXTENDED FAMILY	☐ AUTHORITY FIGURE	☐ BOY/GIRL FRIEND
☐ EX-BOYFRIEND OR EX-GIRLFRIEND	☐ FRIEND	☐ BUSINESS RELATIONSHIP
☐ CRIMINAL RELATIONSHIP	☐ CASUAL ACQUAINTANCE	☐ STRANGER

OCCUPANCY

☐ NOT APPLICABLE	☐ VICTIM IS RESIDENT, ACCUSED NOT KNOWN
☐ VICTIM IS RESIDENT	☐ VICTIM IS NOT RESIDENT, ACCUSED NOT KNOWN
☐ ACCUSED IS RESIDENT	☐ VICTIM AND ACCUSED ARE NOT RESIDENTS

LIVING TOGETHER

☐ UNKNOWN	☐ YES	☐ NO

OFFICER STATUS (ONLY IF OFFICER IS THE VICTIM)

☐ NOT APPLICABLE	☐ POLICE	☐ OTHER PEACE-PUBLIC OFFICER

REMARKS

NARRATIVE:

(TRANSCRIBE VICTIM REPORT NARRATIVE AS DICTATED)

SAULT STE. MARIE POLICE SERVICE
DATA ENTRY GUIDE

WEAPON CAUSING INJURY

☐	NOT APPLICABLE	☐	UNKNOWN	☐	FULLY AUTOMATIC
☐	SAWED-OFF RIFLE OR SHOTGUN	☐	HANDGUN	☐	RIFLE OR SHOTGUN
☐	OTHER FIREARM	☐	KNIFE	☐	OTHER PIERCING OR CUTTING
☐	CLUB	☐	EXPLOSIVES	☐	FIRE
☐	PHYSICAL FORCE	☐	OTHER WEAPON		

LEVEL OF INJURY

☐	NOT APPLICABLE	☐	UNKNOWN	☐	NO INJURIES
☐	MINOR INJURY	☐	MAJOR PHYSICAL INJURY	☐	LOSS OF LIFE

ACCUSED IS (RELATIONSHIP TO VICTIM)

☐	UNKNOWN	☐	SPOUSE	☐	SEP. OR DIVORCED
☐	PARENT	☐	CHILD	☐	OTHER IMMEDIATE
☐	EXTENDED FAMILY	☐	AUTHORITY FIGURE	☐	BOY/GIRL FRIEND
☐	EX-BOYFRIEND OR EX-GIRLFRIEND	☐	FRIEND	☐	BUSINESS RELATIONSHIP
☐	CRIMINAL RELATIONSHIP	☐	CASUAL ACQUAINTANCE	☐	STRANGER

OCCUPANCY

☐	NOT APPLICABLE	☐	VICTIM IS RESIDENT, ACCUSED NOT KNOWN
☐	VICTIM IS RESIDENT	☐	VICTIM IS NOT RESIDENT, ACCUSED NOT KNOWN
☐	ACCUSED IS RESIDENT	☐	VICTIM AND ACCUSED ARE NOT RESIDENTS

LIVING TOGETHER

☐	UNKNOWN	☐	YES	☐	NO

OFFICER STATUS (ONLY IF OFFICER IS THE VICTIM)

☐	NOT APPLICABLE	☐	POLICE	☐	OTHER PEACE-PUBLIC OFFICER

REMARKS

NARRATIVE:

(TRANSCRIBE VICTIM REPORT NARRATIVE AS DICTATED)

SAULT STE. MARIE POLICE SERVICE
DATA ENTRY GUIDE

MISSING PERSON REPORT

OCCURRENCE:	TASK:
AUTHOR: (DICTATING OFFICER)	REPORT TIME: (TIME OF DICTATION)
ENTERED BY: (TRANSCRIBER)	ENTERED TIME: (DEFAULTS T PRESENT TIME)

MISSING PERSON TYPE:

☐	COMPASSIONATE TO LOCATE	☐ ELOPEE
☐	MISSING	☐ YOUNG OFFENDER
☐	OTHER (CLARIFY)	

MISSING BETWEEN:

FROM DATE AND TIME:	TO DATE AND TIME:

PROBABLE REASON:

☐ ABDUCTION BY STRANGER	☐ ACCIDENT
☐ PARENTAL ABDUCTION CUSTODY ORDER	☐ PARENTAL ABDUCTION NO CUSTODY ORDER
☐ RUNAWAY	☐ UNKNOWN
☐ WANDERED OFF/LOST	☐ OTHER (CLARIFY)

MISSING FROM:

☐ CHILD CARE CENTRE	☐ DETENTION CENTRE	☐ DISASTER	☐ FAMILY RES.
☐ FOSTER HOME	☐ SCHOOL	☐ SHOPPING MALL	☐ VAC./TRAVEL
☐ WORK/WORK RELATED	☐ YOUTH CENTRE	☐ OTHER INSTIT.	☐ OTHER (CLARIFY)

HISTORY:

☐ HABITUAL/CHRONIC	☐ REPEAT	☐ FAMILY RESIDENCE

DENTAL:

☐ AVAILABLE, NOT ENTERED	☐ AVAILABLE (PARTIAL), NOT ENTERED
☐ ENTERED	☐ ENTERED (PARTIAL)
☐ NOT REQUIRED	☐ UNAVAILABLE

PHOTO:

☐ AVAILABLE, NOT ENTERED	☐ AVAILABLE (PARTIAL), NOT ENTERED
☐ ENTERED	☐ ENTERED (PARTIAL)
☐ NOT REQUIRED	☐ UNAVAILABLE

XRAY:

☐ AVAILABLE, NOT ENTERED	☐ AVAILABLE (PARTIAL), NOT ENTERED
☐ ENTERED	☐ ENTERED (PARTIAL)
☐ NOT REQUIRED	☐ UNAVAILABLE

Note: DENTAL, PHOTO, XRAY – EACH IS CONSIDERED A WHOLE.

DISABILITIES/DEPENDENCIES:

☐ ALCOHOL	☐ DRUGS
☐ MEDICAL	☐ MENTAL DISABILITY
☐ PHYSICAL DISABILITY	☐ SOLVENT ABUSE
☐ SUICIDE RISK	☐ OTHER (CLARIFY)

PROBABLE DESTINATION:

INSTITUTION	ORDER EXPIRY DATE:

LAST SEEN BY:	LAST SEEN AT:

REMARKS:

NARRATIVE:

(TRANSCRIBE MISSING PERSON REPORT NARRATIVE AS DICTATED)

ENTER DESCRIPTION, MARKS/CLOTHING IN APPROPRIATE TABS

SAULT STE. MARIE POLICE SERVICE
DATA ENTRY GUIDE

MISSING PERSON REPORT

OCCURRENCE:	TASK:
AUTHOR: (DICTATING OFFICER)	REPORT TIME: (TIME OF DICTATION)
ENTERED BY: (TRANSCRIBER)	ENTERED TIME: (DEFAULTS T PRESENT TIME)

MISSING PERSON TYPE:

☐ COMPASSIONATE TO LOCATE	☐ ELOPEE
☐ MISSING	☐ YOUNG OFFENDER
☐ OTHER (CLARIFY)	

MISSING BETWEEN:

FROM DATE AND TIME:	TO DATE AND TIME:

PROBABLE REASON:

☐ ABDUCTION BY STRANGER	☐ ACCIDENT
☐ PARENTAL ABDUCTION CUSTODY ORDER	☐ PARENTAL ABDUCTION NO CUSTODY ORDER
☐ RUNAWAY	☐ UNKNOWN
☐ WANDERED OFF/LOST	☐ OTHER (CLARIFY)

MISSING FROM:

☐ CHILD CARE CENTRE	☐ DETENTION CENTRE	☐ DISASTER	☐ FAMILY RES.
☐ FOSTER HOME	☐ SCHOOL	☐ SHOPPING MALL	☐ VAC./TRAVEL
☐ WORK/WORK RELATED	☐ YOUTH CENTRE	☐ OTHER INSTIT.	☐ OTHER (CLARIFY)

HISTORY:

☐ HABITUAL/CHRONIC	☐ REPEAT	☐ FAMILY RESIDENCE

DENTAL:

☐ AVAILABLE, NOT ENTERED	☐ AVAILABLE (PARTIAL), NOT ENTERED
☐ ENTERED	☐ ENTERED (PARTIAL)
☐ NOT REQUIRED	☐ UNAVAILABLE

PHOTO:

☐ AVAILABLE, NOT ENTERED	☐ AVAILABLE (PARTIAL), NOT ENTERED
☐ ENTERED	☐ ENTERED (PARTIAL)
☐ NOT REQUIRED	☐ UNAVAILABLE

XRAY:

☐ AVAILABLE, NOT ENTERED	☐ AVAILABLE (PARTIAL), NOT ENTERED
☐ ENTERED	☐ ENTERED (PARTIAL)
☐ NOT REQUIRED	☐ UNAVAILABLE

Note: DENTAL, PHOTO, XRAY – EACH IS CONSIDERED A WHOLE.

DISABILITIES/DEPENDENCIES:

☐ ALCOHOL	☐ DRUGS
☐ MEDICAL	☐ MENTAL DISABILITY
☐ PHYSICAL DISABILITY	☐ SOLVENT ABUSE
☐ SUICIDE RISK	☐ OTHER (CLARIFY)

PROBABLE DESTINATION:

INSTITUTION	ORDER EXPIRY DATE:

LAST SEEN BY:	LAST SEEN AT:

REMARKS:

NARRATIVE:

(TRANSCRIBE MISSING PERSON REPORT NARRATIVE AS DICTATED)

ENTER DESCRIPTION, MARKS/CLOTHING IN APPROPRIATE TABS

SAULT STE. MARIE POLICE SERVICE
DATA ENTRY GUIDE

HOMICIDE/SUDDEN DEATH REPORT

OCCURRENCE #:	**TASK #:**
AUTHOR: (DICTATING OFFICER)	**REPORT TIME: (TIME OF DICTATION)**
ENTERED BY: (TRANSCRIBER)	**ENTERED TIME: (DEFAULTS TO PRESENT TIME)**

TYPE:

☐ HOMICIDE	☐ SUDDEN DEATH	☐ SUICIDE	☐ UNKNOWN	☐ OTHER (CLARIFY)
TIME OF DEATH:		**TO:**		
PRONOUNCED DEAD BY:	**TEL #:**		**TIME:**	
CORONER:	**TEL #:**		**TIME:**	
PATHOLOGIST	**TEL #:**		**TIME:**	
POST-MORTEM TIME:				

NEXT OF KIN:

NAME:	**TEL #:**
NOTIFIED BY:	**NOTIFIED TIME:**

BODY:

RELEASED BY:	**RELEASED TIME:**
TAKEN BY:	**TAKEN TIME:**
TAKEN TO:	

ADDITIONAL INFORMATION TO BE ENTERED:

WEAPON:	**MOTIVE:**
CAUSE OF DEATH:	
IDENTITY ESTABLISHED:	
REMARKS:	

NARRATIVE:

(TRANSCRIBE HOMICIDE/SUDDEN DEATH REPORT NARRATIVE AS DICTATED)

SAULT STE. MARIE POLICE SERVICE
DATA ENTRY GUIDE

HOMICIDE/SUDDEN DEATH REPORT

OCCURRENCE #:	**TASK #:**
AUTHOR: (DICTATING OFFICER)	**REPORT TIME: (TIME OF DICTATION)**
ENTERED BY: (TRANSCRIBER)	**ENTERED TIME: (DEFAULTS TO PRESENT TIME)**

TYPE:

☐ HOMICIDE	☐ SUDDEN DEATH	☐ SUICIDE	☐ UNKNOWN	☐ OTHER (CLARIFY)

TIME OF DEATH:		**TO:**	
PRONOUNCED DEAD BY:	**TEL #:**		**TIME:**
CORONER:	**TEL #:**		**TIME:**
PATHOLOGIST	**TEL #:**		**TIME:**
POST-MORTEM TIME:			

NEXT OF KIN:

NAME:	**TEL #:**
NOTIFIED BY:	**NOTIFIED TIME:**

BODY:

RELEASED BY:	**RELEASED TIME:**
TAKEN BY:	**TAKEN TIME:**
TAKEN TO:	

ADDITIONAL INFORMATION TO BE ENTERED:

WEAPON:	**MOTIVE:**
CAUSE OF DEATH:	
IDENTITY ESTABLISHED:	
REMARKS:	

NARRATIVE:

(TRANSCRIBE HOMICIDE/SUDDEN DEATH REPORT NARRATIVE AS DICTATED)

SAULT STE. MARIE POLICE SERVICE
DATA ENTRY GUIDE

WITNESS STATEMENT/WILL SAY

OCCURRENCE:	TASK:
STMT. TAKER: (OFFICER)	STMT. TIME: (TIME STATEMENT TAKEN)
ENTERED BY: (TRANSCRIBER)	ENTERED TIME: (DEFAULTS TO PRESENT TIME)
REMARKS:	

NARRATIVE:

(TRANSCRIBE WITNESS STATEMENT/WILL SAY AS DICTATED)

SAULT STE. MARIE POLICE SERVICE
DATA ENTRY GUIDE

WITNESS STATEMENT/WILL SAY

OCCURRENCE:	TASK:
STMT. TAKER: (OFFICER)	STMT. TIME: (TIME STATEMENT TAKEN)
ENTERED BY: (TRANSCRIBER)	ENTERED TIME: (DEFAULTS TO PRESENT TIME)
REMARKS:	

NARRATIVE:

(TRANSCRIBE WITNESS STATEMENT/WILL SAY AS DICTATED)

SAULT STE. MARIE POLICE SERVICE
POLICE OFFICER GUIDE

S/SGT.	☐
DUTY OFFICER	☐
CERB	☐
SHIFT CLERK	☐

☐ ADULT ☐ YOUNG OFFENDER	☐ MALE ☐ FEMALE	CELL NO. ☐ ☐

ARRESTING OFFICER: BADGE #:

OCCURRENCE ADDRESS:

CHARGES:

OCCURRENCE #:	CHARGE:	OCC DATE:
OCCURRENCE #:	CHARGE:	OCC DATE:
OCCURRENCE #:	CHARGE:	OCC DATE:
OCCURRENCE #:	CHARGE:	OCC DATE:
OCCURRENCE #:	CHARGE:	OCC DATE:

OCCURRENCE ADDRESS TYPE:

☐ UNKNOWN ☐ SINGLE HOME (YARD, DRIVEWAY) ☐ PRIVATE PROPERTY STRUCTURE 　(E.G. DETACHED GARAGE, SHED) ☐ APARTMENT ☐ HOTEL ☐ CAR DEALERSHIP ☐ FINANCIAL INSTITUTION	☐ CONVENIENCE STORE ☐ GAS STATION ☐ SUPERVISED SCHOOL ☐ UNSUPERVISED SCHOOL ☐ COLLEGE ☐ OTHER COMMERCIAL PLACE 　(E.G. OFFICE BLDG, STORE, BAR, 　RESTAURANT)	☐ OTHER NON-COMMERCIAL PLACE 　(E.G. COURTHOUSE, POLICE STN, 　CITY HALL, HOSPITAL, CHURCH) ☐ PARKING LOTS ☐ BUS OR BUS SHELTER ☐ OTHER PUBLIC TRANSPORTATION ☐ STREET OR HIGHWAY ☐ OPEN AREAS

MOST SERIOUS WEAPON USED:

☐ NOT APPLICABLE ☐ UNKNOWN ☐ FULLY AUTOMATIC ☐ SAWED-OFF RIFLE OR SHOTGUN ☐ HANDGUN ☐ RIFLE OR SHOTGUN	☐ OTHER FIREARM ☐ KNIFE ☐ OTHER PIERCING OR CUTTING ☐ CLUB ☐ EXPLOSIVES	☐ FIRE ☐ PHYSICAL FORCE ☐ OTHER WEAPON ☐ THREAT ☐ NO WEAPON

WEAPON STATUS:

☐ NOT APPLICABLE	☐ UNKNOWN	☐ REAL	☐ FACSIMILE

NAME TYPE:

☐ PRIMARY ☐ ALIAS	☐ NICKNAME ☐ MAIDEN NAME	☐ VARIANT ☐ OTHER	☐ ACRONYM ☐ LEGAL	☐ OPERATING

ACCUSED PERSON:

SURNAME:	G1:	G2:	G3:

SEX:	DOB:	(OR AGE):

NAME TYPE:

☐ PRIMARY ☐ ALIAS	☐ NICKNAME ☐ MAIDEN NAM	☐ VARIANT ☐ OTHER	☐ ACRONYM ☐ LEGAL	☐ OPERATING

ALIAS:	NICKNAME:

SAULT STE. MARIE POLICE SERVICE
POLICE OFFICER GUIDE

S/SGT.	☐
DUTY OFFICER	☐
CERB	☐
SHIFT CLERK	☐

☐ ADULT ☐ YOUNG OFFENDER	☐ MALE ☐ FEMALE	CELL NO. ☐ ☐

ARRESTING OFFICER: **BADGE #:**

OCCURRENCE ADDRESS:

CHARGES:

OCCURRENCE #:	CHARGE:	OCC DATE:
OCCURRENCE #:	CHARGE:	OCC DATE:
OCCURRENCE #:	CHARGE:	OCC DATE:
OCCURRENCE #:	CHARGE:	OCC DATE:
OCCURRENCE #:	CHARGE:	OCC DATE:

OCCURRENCE ADDRESS TYPE:

☐ UNKNOWN
☐ SINGLE HOME (YARD, DRIVEWAY)
☐ PRIVATE PROPERTY STRUCTURE
  (E.G. DETACHED GARAGE, SHED)
☐ APARTMENT
☐ HOTEL
☐ CAR DEALERSHIP
☐ FINANCIAL INSTITUTION
	☐ CONVENIENCE STORE
	☐ GAS STATION
	☐ SUPERVISED SCHOOL
	☐ UNSUPERVISED SCHOOL
	☐ COLLEGE
	☐ OTHER COMMERCIAL PLACE
	  (E.G. OFFICE BLDG, STORE, BAR,
	  RESTAURANT)
		☐ OTHER NON-COMMERCIAL PLACE
		  (E.G. COURTHOUSE, POLICE STN,
		  CITY HALL, HOSPITAL, CHURCH)
		☐ PARKING LOTS
		☐ BUS OR BUS SHELTER
		☐ OTHER PUBLIC TRANSPORTATION
		☐ STREET OR HIGHWAY
		☐ OPEN AREAS

MOST SERIOUS WEAPON USED:

☐ NOT APPLICABLE ☐ UNKNOWN ☐ FULLY AUTOMATIC ☐ SAWED-OFF RIFLE OR SHOTGUN ☐ HANDGUN ☐ RIFLE OR SHOTGUN	☐ OTHER FIREARM ☐ KNIFE ☐ OTHER PIERCING OR CUTTING ☐ CLUB ☐ EXPLOSIVES	☐ FIRE ☐ PHYSICAL FORCE ☐ OTHER WEAPON ☐ THREAT ☐ NO WEAPON

WEAPON STATUS:

☐ NOT APPLICABLE	☐ UNKNOWN	☐ REAL	☐ FACSIMILE

NAME TYPE:

☐ PRIMARY ☐ ALIAS	☐ NICKNAME ☐ MAIDEN NAME	☐ VARIANT ☐ OTHER	☐ ACRONYM ☐ LEGAL	☐	OPERATING

ACCUSED PERSON:

SURNAME:	G1:	G2:	G3:

SEX:	DOB:	(OR AGE):

NAME TYPE:

☐ PRIMARY ☐ ALIAS	☐ NICKNAME ☐ MAIDEN NAM	☐ VARIANT ☐ OTHER	☐ ACRONYM ☐ LEGAL	☐	OPERATING

ALIAS:	NICKNAME:

Index